EMPOWERING ELECTRICITY

The Sustainability and the Environment series provides a comprehensive, independent, and critical evaluation of environmental and sustainability issues affecting Canada and the world today.

Anthony Scott, John Robinson, and David Cohen, eds., *Managing Natural Resources in British Columbia: Markets, Regulations, and Sustainable Development*

John B. Robinson, *Life in 2030: Exploring a Sustainable Future for Canada*

Ann Dale and John B. Robinson, eds., *Achieving Sustainable Development*

John T. Pierce and Ann Dale, eds., *Communities, Development, and Sustainability across Canada*

Robert F. Woollard and Aleck Ostry, eds., *Fatal Consumption: Rethinking Sustainable Development*

Ann Dale, *At the Edge: Sustainable Development in the 21st Century*

Mark Jaccard, John Nyboer, and Bryn Sadownik, *The Cost of Climate Policy*

Glen Filson, ed., *Intensive Agriculture and Sustainability: A Farming Systems Analysis*

Mike Carr, *Bioregionalism and Civil Society: Democratic Challenges to Corporate Globalism*

Ann Dale and Jenny Onyx, eds., *A Dynamic Balance: Social Capital and Sustainable Community Development*

Ray Côté, James Tansey, and Ann Dale, eds., *Linking Industry and Ecology: A Question of Design*

Glen Toner, ed., *Sustainable Production: Building Canadian Capacity*

Ellen Wall, Barry Smit, and Johanna Wandel, eds., *Farming in a Changing Climate: Agricultural Adaptation in Canada*

Derek Armitage, Fikret Berkes, Nancy Doubleday, eds., *Adaptive Co-Management: Collaboration, Learning, and Multi-Level Governance*

Martin K. Luckert, David Haley, and George Hoberg, *Policies for Sustainably Managing Canada's Forests: Tenure, Stumpage Fees, and Forest Practices*

Rod MacRae and Elisabeth Abergel, eds., *Health and Sustainability in the Canadian Food System: Advocacy and Opportunity for Civil Society*

Sara Teitelbaum, ed., *Community Forestry in Canada: Lessons from Policy and Practice*

SUSTAINABILITY
AND THE
ENVIRONMENT

EMPOWERING ELECTRICITY

CO-OPERATIVES, SUSTAINABILITY, AND POWER SECTOR REFORM IN CANADA

Julie L. MacArthur

UBCPress · Vancouver · Toronto

25 24 23 22 21 20 19 18 17 16   5 4 3 2 1

Printed in Canada on FSC-certified ancient-forest-free paper (100% post-consumer recycled) that is processed chlorine- and acid-free.

Library and Archives Canada Cataloguing in Publication

MacArthur, Julie L., author
    Empowering electricity : co-operatives, sustainability, and power sector reform in Canada / Julie L. MacArthur.

(Sustainability and the environment)
Includes bibliographical references and index.
Issued in print and electronic formats.
ISBN 978-0-7748-3143-7 (hbk). – ISBN 978-0-7748-3145-1 (pdf). –
ISBN 978-0-7748-3146-8 (epub). – ISBN 978-0-7748-3147-5 (mobi)

    1. Electric cooperatives – Canada. 2. Electric power production – Environmental aspects – Canada. 3. Electric power – Canada – Citizen participation. 4. Community development – Canada. 5. Renewable energy sources – Canada. 6. Sustainable development – Canada. 7. Consumer cooperatives – Canada. 8. Energy policy – Canada. I. Title. II. Series: Sustainability and the environment

| HD9685.C32M27 2016 | 334.68133379320971 | C2015-908712-0 |
| | | C2015-908713-9 |

Canada

UBC Press gratefully acknowledges the financial support for our publishing program of the Government of Canada (through the Canada Book Fund), the Canada Council for the Arts, and the British Columbia Arts Council.

This book has been published with the help of a grant from the Canadian Federation for the Humanities and Social Sciences, through the Awards to Scholarly Publications Program, using funds provided by the Social Sciences and Humanities Research Council of Canada.

Printed and bound in Canada by Friesens
Set in Segoe and Warnock by Apex CoVantage, LLC
Copy editor: Judy Phillips
Proofreader: Francis Chow
Cover designer: Martyn Schmoll

UBC Press
The University of British Columbia
2029 West Mall
Vancouver, BC V6T 1Z2
www.ubcpress.ca

Contents

List of Tables / vii

Preface and Acknowledgments / ix

Abbreviations / xiii

1 A Climate for Change / 1

2 Governing Sustainability: From Crisis to Empowerment / 28

3 Co-operatives in Canadian Political Economy / 45

4 International Forces for Power-Sector Restructuring / 67

5 Continental, Private, and Green(er)? Canadian Electricity Restructuring / 86

6 Electricity Co-operatives: The Power of Public Policy / 116

7 Off the Ground and on the Grid: New Electricity Co-operative Development / 139

8 Co-operative Networks and the Politics of Community Power / 165

9 Empowering Electricity / 183

Appendices

1 List of Personal Communications / 202

2 Co-operative Policies and Programs across Canada / 207

3 Top Ten Nonfinancial Co-ops, 2010 / 210

4 International Co-operative Principles / 211

5 Major Proposed IPL Transmission Lines, 2014 / 213

6 Ontario FIT and RESOP Prices / 215

Notes / 218

Glossary / 224

References / 228

Index / 252

Tables

1.1 Comparison of international wind-generation ownership structures / 19

2.1 Countervailing power / 42

2.2 Framework for assessing co-operative potential / 43

3.1 Types of co-operatives / 49

3.2 Co-operatives by province and membership, 2010 / 50

3.3 Co-operative and business comparison / 52

3.4 Co-operative and business survival rate in Quebec / 55

3.5 Provinces with specific co-operative government agencies / 63

4.1 Electricity governance regimes / 71

4.2 International electricity exports by province, 2012 / 79

5.1 Majority ownership and fuel source by province, 2012 / 88

5.2 Residential electricity price and fuel source, 2015 / 89

5.3 Canadian electricity generation by source, 2012 / 90

5.4 Life-cycle assessment of GHG emissions (kt CO_2 eq./TWh) / 90

5.5 Electricity generation by province and source, 2012 / 91

5.6 Federal and provincial installed capacity by ownership, 2011 / 92

5.7 Initial provincial electricity restructuring policies / 94

5.8 Changing public and private share of installed capacity, 2000–11 / 96

5.9 IPP purchases in British Columbia, 2000–13 / 101

5.10 Installed wind capacities by province or territory and ownership, 2011 / 103

5.11 Total installed wind capacity in Canada, 1995–2014 / 104

5.12 Sample Ontario RESOP and FIT rates, 2006–14 / 109

5.13 Residential electricity rates in Canadian cities (¢/kWh) / 113

6.1 Electricity co-operatives by type incorporated, 1940–2013 / 118

6.2 Electricity co-operatives by province and period incorporated, 1940–2013 / 119

6.3 Electricity co-operatives by province and type incorporated, 1990–2013 / 124

6.4 Provincial community power policies, 2014 / 130

7.1 Status of electricity co-operatives in Canada, 2013 / 142

7.2 Selected electricity generation co-operatives by province and structure, 2013 / 145

Preface and Acknowledgments

In November 2015, world leaders met in Paris for the twenty-first Conference of the Parties to the United Nations Framework Convention on Climate Change. The existential threat climate change poses to human civilization requires a universal and binding agreement. However, these high-level meetings are now accompanied by widespread skepticism that, despite new pledges, emissions will continue to increase as policy makers remain divided over the science, politics, and economics of climate policy. Canada, in particular, has long been a laggard at these meetings, despite our relative wealth and high per-capita contribution to greenhouse gas emissions. Our response to this complexity and the scale of the challenge cannot be resignation or despair; the climate crisis is too significant to put in the "too hard" basket of policy problems. Even with the electoral winds changing in Canada and more climate-active governments in Ottawa and Edmonton, the challenges facing us are multilayered. International regimes, for example, form just the tip of a very complex iceberg of climate governance, albeit a very important tip. As a result, researchers like me have increasingly turned to investigating the power of substate and nonstate actors in provoking and sustaining climate action.

The roots of *Empowering Electricity* trace back to my time at Simon Fraser University. I was interested in how citizens in countries with abundant natural and financial resources can move forward with climate action in the face of ineffective or even hostile political leadership. I realized early on that issues of climate policy implementation, particularly in renewable energy, are deeply connected to persistent political debates over economic power and ownership, appropriate governing scales, and economic versus representative democracy. New energy policies and projects routinely come into conflict with provincial, regional, and local priorities and are accompanied by significant economic and biophysical impacts. Facilities have to be in someone's backyard, and the economic benefits have historically accrued to centralized public coffers or – increasingly in Canada – to multinational private ones. These challenges spurred a desire to explore climate action innovations from the bottom up, by focusing on the role that community associations – co-operatives in this instance – play in the renewable electricity sector.

When I began the project in 2006, "community power" and "community energy" were not common terms in Canada. A decade later, the growth and development in community power/energy has been remarkable thanks to a combination of international policy learning and local entrepreneurial efforts in key jurisdictions. These changes are potentially significant as we sit at a crucial juncture in the development of the Canadian electricity systems. The pressure to upgrade, innovate, and modernize arises from the urgency of climate action as well as from the potential returns from growth in the renewables industry. It is not that Canadians are ill-equipped technically or economically for radical reductions in fossil fuel consumption for heating, transport, and other vital activities; low-emissions technologies exist, and their costs are falling rapidly. Rather, the issue is that any significant shift away from incumbent sources and industries is accompanied by economic and social costs, so political minefields abound.

Research into alternative economies, both in scale and purpose, is flourishing today, and many observers argue that institutional changes may help to address some of the "wicked" problems associated with sustainable transitions. These include perennial issues that political scientists are well placed to address: challenges of democratic engagement, perverse incentives, and informational asymmetries. In the early stages of research, Elinor Ostrom's work on collective action, and commons management in particular, piqued my curiosity about the role of institutional structure on behavioural change. From there, the task was set to track the number, nature, and power of these actors in the renewable energy sector across Canada. This research led me to critical political economy accounts of co-operatives and the social economy that challenge simplistic understandings of community, empowerment, and sustainability. The stakes in the debate over the potential of local initiatives are significant: on the one hand, community energy might provide mechanisms for more just and sustainable transitions; on the other hand, it may serve to underpin deeply unsustainable norms.

Consider the following illustrative example of the challenges facing us. A world away from global climate summit meetings, in the Peace region of British Columbia, fierce local opposition to a proposed hydro-power dam (Site C) has intensified following the province's December 2014 approval of project construction. The BC government has argued that the benefits in terms of new low-emissions power generation will help meet climate targets and provide much-needed reliability in the electricity system. Local residents of Fort St. John, together with First Nations and rural communities, object to the environmental impacts of flooding thousands of hectares along a fertile river valley and construction of a third dam along the Peace River. The costs and benefits of new projects, from Site C to Ontario's nuclear installations, raise pressing questions. Which new energy developments are necessary? When and where should considerations of sufficiency overwhelm those of efficiency? What role can and should those living closest to projects play?

These questions extend far beyond the development of one particular dam, or pipeline, or coal-plant to deeply political issues of economic and environmental justice. They also highlight practical challenges associated with climate action and policy design. Although a substantial literature explores the political economy of renewable electricity and climate policies, very little scholarly work focuses on the impact of community and co-operative power generation in Canada. Texts that address electricity regimes and energy policy often ignore nontraditional actors, and social economy work in the country rarely examines the complexities of development in energy and electricity systems. This book seeks to fill this gap. It is based on in-depth empirical research and provides a current overview of Canadian renewable electricity co-operatives, as well as a contextually and historically grounded analysis. My hope is that the account of electricity co-operative developments presented here stimulates debate and discussion for academics, practitioners, and policy makers alike.

As with any research of this scope, I could not do it alone. It is difficult to overstate my gratitude to the scholars, practitioners, friends, and family members who have shaped my intellectual life over the past decade. The writing was a solitary process, but the project itself most certainly was not. My research would not have been possible without the generosity of the many interviewees and academics across the country who shared their experiences and sometimes even their homes, often with warmth and enthusiasm. I am extremely grateful for their time, insights, and inspiring work. My ongoing collaborations with the BC-Alberta Social Economy Research Alliance provided both intellectual and financial nourishment, particularly in the Alberta sections, so thank you to Mike Lewis, Mike Gismondi, Noel Keough, Stuart Wulff, Sean Connelly, Lena Soots, and Paul Cabaj, among many others.

Empowering Electricity owes much to the supportive colleagues I've been fortunate enough to have at the University of Auckland and Simon Fraser University. There are far too many for a comprehensive list, so what follows below is a start. I completed the manuscript edits in my first two years of a new job at the University of Auckland, so the encouragement of my colleagues, particularly Anita Lacey, Stephen Noakes, Tom Gregory, and Annie Bartos, to keep motivated each semester as the marking and admin descended has been truly invaluable. I'd like to acknowledge too the excellent work of my research assistant Claudia Gonnelli in helping me with updated statistics and final edits and Kate Kennedy in editing an early version of the work. At Simon Fraser University, Marjorie Griffin Cohen, Stephen McBride, Genevieve Fuji Johnson, David Laycock, Heather Whiteside, and Joshua Newman provided intellectual support for what proved to be a very challenging and multifaceted project. I would also like to thank J.J. McMurtry for his thorough engagement with my manuscript and his continued work in researching community renewables across Canada with Judith Lipp as part of the People, Power, Profit initiative.

This research would not have been possible without the financial support of the Social Sciences and Humanities Research Council through the Awards to Scholarly Publications Program, as well as its support of the BC-Alberta Social Economy Research Alliance (BALTA) partnership. Simon Fraser University and the University of Auckland have provided financial support for the research and writing of this manuscript. The publication team at UBC Press, Randy Schmidt, Ann Macklem, Judy Phillips, and Nadine Pedersen have been unfailingly patient, shepherding me through the book publication process and reviewing the manuscript with a careful eye. Any errors that remain are, of course, my own.

Finally, to my parents, Ed and Joni, and to my husband, Francis: thank you for empowering me to do what I love.

Abbreviations

BCSEA	British Columbia Sustainable Energy Association
CANWEA	Canadian Wind Energy Association
CAREA	Central Alberta Rural Electrification Association
CCS	carbon capture and storage
CEDIF	Community Economic Development Investment Funds
CDR	co-opératives de développement régional
CHP	combined heat and power
COMFIT	community feed-in tariff
CPF	Community Power Fund
CSP	cross-sector partnerships
EPG	empowered participatory governance
FERC	Federal Energy Regulatory Commission
FIT	feed-in tariff
GATT	General Agreement on Tariffs and Trade
GHG	greenhouse gas
GWh	gigawatt hour
IESO	Independent Electricity System Operator
IOU	investor-owned utility
IPL	international power line
IPP	independent power producer
kWh	kilowatt hour
LREC	Lamèque Renewable Energy Co-operative

MWh	megawatt hour
NAFTA	North American Free Trade Agreement
NERC	North American Electric Reliability Corporation
NIMBY	not in my backyard
OATT	open access transmission tariff
OSEA	Ontario Sustainable Energy Association
PEC	Peace Energy Co-operative
PV	photovoltaics
REA	rural electrification association
RESOP	Renewable Energy Standard Offer Program
TREC	Toronto Renewable Energy Co-operative
TWh	terawatt hour

EMPOWERING ELECTRICITY

1

A Climate for Change

What are the core institutions of a sustainable electricity system? Despite important differences, both state-led and market-led modes of governance have generally sacrificed ecological considerations for economic priorities and disconnected the majority of citizens from decision making regarding resource allocation. These detrimental choices are significant in Canada, where mounting evidence of global climate change is accompanied by some of the highest per capita greenhouse gas footprints in the world (Environment Canada 2013). This is partially because of persistent reliance on fossil fuel extraction for domestic energy use as well as for export. However, radical transformations of electricity governance are currently underway in some countries, altering how power is generated and distributed. These transformations include an expansion in actors beyond the traditional power sector – centred on public and private utilities – and toward a network of alternative players: electricity co-operatives. There were more than two hundred of these locally owned and democratically structured organizations operating across Canada in 2015.

In many cases, electricity co-operative development is accompanied by enthusiasm that they may represent a greener, more locally based and democratic form of electricity ownership. New co-operatives are nearly all focused on developing cleaner and more sustainable sources of power: wind, solar, biomass, tidal, and hydro. They differ in important ways from centralized-state- and private-shareholder-controlled firms, so a democratization of the electricity sector by co-operatives may be desirable. For example, critics of liberal democracy's close connection with capitalism, such as C.B. MacPherson, have long argued that democracy cannot flourish without a restructuring of economic relationships and a socialization of ownership (Macpherson 1973, 1977). Substantive democracy, in this view, requires a broadening from formal political institutions into economic and, more recently, environmental ones (Adkin 2009); it includes deepening democracy via promotion of institutions that enhance both participation and deliberation (Fung and Wright 2003; Pateman 1988; Wright 2010). In recent years, the intractable and "wicked" nature of many environmental challenges has also strengthened calls for new forms of participatory democratic governance (Blay-Palmer 2011; Catney, Dobson, et al. 2013; Krupa, Galbraith, and Burch 2013).

This book systematically explores the development of Canadian electricity co-operatives as they relate to larger debates over renewable electricity policy and sustainable governance transitions. My aim is to assess the potential of these organizations to contribute to a more sustainable energy future in Canada, given the pressing challenge that global climate change presents and the significant changes taking place in electricity systems (Gillis 2014; REN21 2014). Policy changes play a key role in transforming governance systems, affecting both human behaviour and environmental outcomes (Andersson and Ostrom 2008; Doern and Gattinger 2003; Steurer 2013). Thus, I am particularly interested in the role that state policy changes play in directing the nature and direction of community and co-operative players in the energy sector. Co-operatives are often placed outside (and between) the state and market when they are, in fact, permeated by both (McMurtry 2010, 6–13). An explicit focus situating co-operative development within changing electricity-policy regimes can illuminate how they shape and are in turn shaped by structural and sectoral context.

Detailed examination of the changing electricity sector illuminates the context of co-operative development; it also provides a useful contrasting backdrop for assessing the distinctiveness and potential of electricity co-operatives. For example, while most co-operative electricity organizations are relatively small in size, in Ontario they have joined with other actors in the community power sector to influence provincial electricity policies (Etcheverry 2013). So, in one sense they "punch above their weight." However, they are embedded within much larger processes of state restructuring, where the role they play in the Canadian energy system is complex and sometimes contradictory. Co-operatives may be well positioned to help particular communities with particular challenges, but in a sector such as electricity, sustainability depends on political economy forces well beyond the local level. Emergent forms of electricity governance – interactions between the public and private actors that regulate and hold power at multiple levels beneath, within, between, and above states – are thus both theoretically and empirically significant.

"Power," in the context of this book, holds a double meaning. There is the physical electric power that is generated but also the power to govern, to control what gets produced, where, when, and for whom. Political scientists have traditionally focused their attention on power: who has it, how it is used, whether it is institutionally embedded, and – importantly – whether the exercise of power is sanctioned or legitimized by the polity, often through a variety of democratic practices. These perspectives bring to light the embedded relationship between political and economic power: how power is distributed and exercised not only in traditional areas of the economy, such as finance, but how economic power shapes access to mechanisms of governance and decision making. The importance of analyzing these changing forms of power is not merely in describing the world; rather, it is in contributing to transformations toward both participatory democracy and environmental sustainability. This follows Marx's ([1845] 1974) urging for us not just to study the world but also to *change it*. This goal raises the question, what might make for a substantial contribution in these directions and how would we know?

In Denmark and Germany, co-operative and community energy initiatives have proved successful in broadening energy ownership and facilitating rapid new renewable-energy development (Debor 2014; Gipe 2007; Lauersen 2008; Meyer 2007). Electricity co-operatives generate electricity, manage local distribution systems, and provide energy retail and education services across the provinces and states respectively. Co-operatives, as social enterprises, are private firms distinguished from conventional shareholder-owned corporations by a (relatively) democratic corporate structure and by subscription to a set of seven core principles loosely corresponding to the popular slogan "People before profit."[1] The structure can be more democratic than shareholder-owned firms in the sense that the co-operative's owners are project stakeholders, either service users or producers. Together, these factors form what co-operative theorists and practitioners refer to as the "co-operative difference" (Gossen 1975; MacPherson 2008). Seven core co-operative principles are set out by the International Co-operative Alliance (ICA 2010):

1 Voluntary and open membership
2 Democratic member control
3 Member economic participation
4 Autonomy and independence
5 Education, training, and information
6 Co-operation among co-operatives
7 Concern for community.

The systemic contribution of co-operatives today requires careful empirical assessment. In terms of both electricity assets and access to finance, co-operatives often lack the ability to compete with private-sector developers. A further limitation is that, in practice, the degree of democratic control of the organization varies a great deal from co-operative to co-operative. In addition, "local" and "community" are sometimes idealized in the literature on co-operatives and environmental sustainability in ways that don't always hold up under empirical (or theoretical) scrutiny (Carter 1996; Lionais and Johnstone 2010). That these electricity co-operatives go beyond theoretical contributions to form the kind of locally embedded and democratic alternative in practice is, however, important for the development of sustainable electricity futures in Canada. If electricity co-operatives do provide a significant alternative, their practical strength is contingent on an ability both to succeed within and to transform current institutions and norms. Transformational change may, for example, vary based on development in specific settings (urban or rural), provinces, or business areas (such as generation or distribution) or on the specific motivations of key actors in their start-up phase and their willingness to take on broader issues of environmental justice. The following questions thus emerge, and animate the subsequent chapters in this book:

• Where and in what forms are electricity co-operatives developing in Canada?
• What advantages and/or disadvantages does this organizational form have over more traditional corporate forms?
• Why are new co-operatives in this sector experiencing resurgence in some provinces and not others?

- What role do these co-operatives play in shaping public policy and in supporting or challenging different modes of governance?
- What particular challenges does the electricity sector present for co-operative development, if any?
- How do these organizations interface with radical movements for economic democracy and environmental sustainability?

In order to provide insight into these questions, I examine public policy documents, renewable energy-sector publications, and government electricity statistics covering a period from 1940 to 2014. More than fifty interviews with community electricity developers, co-operative associations, and policy makers, from 2009 to 2013, also form an important source of data; these communications, detailed in Appendix 1 of this book, provided crucial insight into the intention and interpretation of both co-operative project development and policy designs.[2] These interviews generated insights into the tensions and conflicts behind the scenes in a way that isn't typically represented in the literature on community energy projects. Electricity co-operatives were identified primarily by name and activity from provincial and national association publications, provincial co-operative registries where available, and a database of co-operatives formerly managed by the federal Co-operatives Secretariat.

Electricity Co-operatives in Canada: Why Now?

Co-operatives have historically arisen in response to crises and in some cases have demonstrated effective mechanisms for community development, empowerment, and economic democracy (MacPherson 2009). Co-operatives in the electricity sector today are developing because of an interrelated set of social, economic, and environmental challenges – a triple crisis – driven by increasingly market-based governance arrangements and a modern industrial system that fails to adequately account for human and natural worth.[3] The triple crisis is an empirical description of mutually reinforcing linkages between ecosystem breakdown, democratic disempowerment, and an economic system reliant on limitless growth (Daly 1996; Johnston, Gismondi, and Goodman 2006b; Kovel 2007; Panitch and Leys 2006). In practical terms, the triple crisis means that addressing the issue of persistent poverty requires enhancing participation and empowerment, and that dealing with environmental degradation requires more equitable distribution of political and economic power. For Sen (1999), exploitation of human and natural resources can erode citizens' capabilities, reducing their effectiveness in responding to complex challenges. Analyses of issues of ownership, participation, and power in any sustainable transition are thus crucial (Albert 2003; Burkett 2006; Faber 2008). Theorists of participatory governance have also argued that empowering local citizens and democratizing economic institutions can lead to improved environmental (Ostrom 1990) and social justice outcomes.

New Renewables and the Climate Crisis

Global climate change is the most pressing and significant manifestation of the triple crisis. Current patterns of production and consumption are dependent on fossil fuel–based energy that provides a high energy return on investment but also correspondingly high greenhouse gas (GHG) emissions (Homer-Dixon 2009). The scientific consensus on climate change, however, is increasingly clear. According to recent reports from the Intergovernmental Panel on Climate Change (IPCC), warming of the climate system is unequivocal. In some models, this warming of 0.8 degrees Celsius already above preindustrial times is projected to increase to 2 to 3 degrees Celsius by 2050 (United Nations 2013). Current climate changes have led to impacts such as ocean acidification, rising sea levels, more intense droughts, and extreme weather events. In their summary report of the physical science of climate change, IPCC scientists argue that "the atmospheric concentrations of carbon dioxide, methane, and nitrous oxide have increased to levels unprecedented in at least the last 800,000 years. Carbon dioxide concentrations have increased 40% since pre-industrial times, primarily from fossil fuel emissions and secondarily from net land use change emissions. The ocean has absorbed about 30 percent of the emitted anthropogenic carbon dioxide, causing ocean acidification" (Stocker et al. 2013, 11). They also argue that "continued emission of greenhouse gases will cause further warming and long-lasting changes in all components of the climate system, increasing the likelihood of severe, pervasive and irreversible impacts for people and ecosystems" (Evans 2014).

Societies unwilling or unable to reduce the anthropogenic climate-change drivers will clearly face substantially increased future socioeconomic costs from a failure to act (IPCC 2011). Importantly, small, vulnerable societies will be significantly impacted regardless of their own mitigation efforts; they are reliant on wealthy resource-intensive nations like Canada and the United States to act responsibly.

Energy services form a fundamental basis for meeting basic human needs for food, shelter, transportation, communication, and development. Fossil fuels such as coal, oil, and gas for energy services has historically dominated energy supply. This has led directly to postindustrial increases in global carbon dioxide emissions. Demand for energy services is predicted to increase significantly – in 2012, 1.3 billion people were still without access to electricity – so the sources and structures developed in this sector are central to solving the climate crisis (International Labour Organization 2013). Renewable energy (RE) has a significant role to play in mitigating GHG emissions growth. A switch to new renewable-energy systems "if implemented properly, contributes to social and economic development, energy access, a secure energy supply, and reducing negative impacts on the environment and health" (REN21 2014, 7). However, the ultimate mitigation potential depends significantly on the fuels and sources displaced by new renewables as well as the specific technologies employed (IPCC 2011, 22).

Renewable energy transitions clearly form a vital piece of the climate change puzzle. How, why, and where energy transitions take place at multiple levels is thus important. Although renewable energy technologies have increased rapidly over the past decade – nearly doubling from 2004 to 2012 – their widespread adoption at the levels required for mitigating climate change necessitates strong targeted policy action (REN21 2009, 2014). In Germany and Denmark, bold policy shifts have resulted in 100 percent renewable energy regions as well as in world-leading policies for renewable heat (as opposed to electricity) supply and a significant role for community actors (International Labour Organization 2013). Germany is currently in the process of phasing out its nuclear power generation by 2022 and increasing its share of renewable energy generation to 30 percent by 2025 as part of its *energiewende* or "energy transition" (Gillis 2014).

Opposition to new renewable-energy policies remains a significant challenge, however, as new actors contest for scarce public funds and access to aging and stressed electrical grids. Indeed, even though new sources are being developed at increasing rates, "enormous subsidies for fossil fuels and nuclear power persist, and they continue to vastly outweigh financial incentives for renewables" (REN21 2014, 104). In addition to direct subsidies, costs of air pollution and its health effects, together with the costs of climate adaptation and mitigation, are not internalized.

Severe informational asymmetries obscure the real costs of production and consumption. It is unlikely that these asymmetries will self-correct, because social and environmental costs are externalized to geographically disparate communities across the globe, often ones with poor political representation and less economic power. This imbalance severs crucial eco-social feedback loops that could (and should) mitigate self-destructive practices, further reinforcing the triple crisis (Ostrom 1990; Princen, Maniates, and Conca 2002). A complex pattern emerges, wherein the relative balance of costs and rents shifts to favour private accumulation over public control. In this way, economic processes continue to degrade the natural environment as well as contribute to disempowering local citizens. A key challenge going forward is thus to identify institutional forms capable of shifting energy governance in a more sustainable direction (Hahnel 2007; Stephenson et al. 2010).

In the Canadian energy sector in particular, the interrelationships between policy actors and incumbent nuclear and fossil fuel industries are an important challenge facing new renewables-policy entrepreneurs (Bratt 2012; Durant 2009; Durant and Johnson 2009; Rowlands 2007). Canada is an energy-rich country and one of the largest producers and exporters of oil, natural gas, coal, uranium, and hydroelectricity in the world (Natural Resources Canada 2008). With some of the highest per capita GHG and carbon dioxide (CO_2) emissions in the world (Homer-Dixon 2007; Paehlke 2008), Canada has both an ethical responsibility and the resource capacity to address climate change.

But policy responses have been slow. Canadian citizens are confronting record levels of income inequality and political disengagement (Pilon 2001; United Nations 2010). As Teeple (2000, 3) argues, this is in part because "the idea that politics determines national policies has gradually dissipated, and in its place has come the open assertion that economics is the deciding factor in more and more aspects of society." The democratic legitimacy of traditional sites of collective action has been eroded through decades of policy shifts hollowing out state agencies and shifting power to market-based actors. These reforms are part of a broader international project of politico-economic restructuring that draws heavily from both neoclassical and Austrian economic thought – referred to in this book as neoliberalism (McBride 2005; Panitch 2007).[4]

The strengthening of neoliberal ideology in Canada over the past three decades has led to the increased marketization and commodification of key natural resources, from British Columbia's rivers to windy coastal sites in Quebec's Gaspé Peninsula (Byrne, Toly, and Glover 2006; Doern and Gattinger 2003). This takes place, for example, via the restructuring of power sectors to facilitate private ownership of new renewable electricity generation. Ultimately, this represents a shift in the mode of socioeconomic governance wherein the normative ideals highlight the virtues and benefits of private-sector growth, and the policy practice cedes ownership and command-and-control regulation in favour of voluntary, marketized, and networked forms of governance.

The triple crisis in Canada has prompted a search for alternative sources of community power: power in the sense of electric power through greener sources, and power in the sense of more democratic and participatory institutions and forms of governance. Co-operatives are part of this broader community power sector, which includes a wide range of actors with diverse organizations and motivations attempting to develop new renewables. Reforming electricity generation, particularly via the development of renewables like wind, solar, and tidal power in coal-reliant provinces (Alberta, Ontario, Saskatchewan, Nova Scotia), may play an important part in the transition to a more sustainable future. However, Canada's deep integration within a North American economy (through, for example, the North American Free Trade Agreement and the World Trade Organization), and an increase in private-sector ownership of new electricity generation, are leading to an ever-increasing erosion of the public sector's share of generating assets.

Electricity-Sector Reform: Power to the Private?
Electricity co-operatives have re-emerged in Canada in the midst of great change in the power sector. Provincial governments across the country have largely engaged in piecemeal restructuring of electricity systems in order to increase participation by private companies, particularly for new renewable generation.[5] Restructuring is occurring because of three key drivers. The first is the influence of pro-market reforms on the public ownership of electricity production in Canada. Provincial reforms have resulted in the privatization of new energy production and some aspects of traditional government-owned utilities. The second driver, deeply related to the first, is the expansion of continental power grids that facilitate generation for export and are regulated by US-based bodies such as the Federal Energy Regulatory Commission and the North American Electric Reliability Corporation. The third driver is the rise of environmental issues as a focus for government action and the resulting move toward new renewable-electricity generation ("green power"). These three drivers are not easily separable, and their confluence has allowed the green power movement to make inroads into territory formerly the realm of Crown corporations, with important implications for participation and power alike.

Canada's electricity resources are vast and lucrative. Unlike oil, electricity in Canada remains primarily in the public sector (discussed in Chapter 4). This is changing, however, as provinces incrementally open markets to private actors for new renewable-power generation. For the last twenty years, provincial and federal governments across the country have been steadily orienting away from nationalism and public control, and toward increased private ownership and continentalism (Calvert 2007; CCPA, Parkland Institute, and Polaris Institute 2006; Cohen 2007). These provincial changes are part of a broader project of power-sector restructuring around the world (Beder 2003) wherein nearly one hundred countries have privatized their electrical utilities since the 1990s. In Canada, these developments are taking place in unique (often piecemeal) ways when compared with other states, since many provincial power sectors are still structured around public and often hydro-based utilities. These trends shape both Canada's distribution of wealth and its citizens' ability to address the pressing and interrelated social, economic, and environmental challenges confronting the country. Chapters 4 and 5 explore these international and domestic processes in more detail.

The current push to increase private-sector access to the remaining public aspects of electricity in Canada exists despite the growing recognition of the costs of privatization and deregulation of resources elsewhere and the importance of strong state action on climate change (Cohen 2006b; Doern and Gattinger 2003; Stocker et al. 2013). The social, economic, and environmental outcomes of investor-owned corporate control continue to be questioned. This is, in part, because profit-based incentive structures and lack of local participatory engagement in governance often lead to socially and environmentally damaging outcomes (Dryzek 1992; Faber 2008; Fitzpatrick 2014; O'Connor 1994). Provinces are ceding public control of critical new assets and are increasingly reduced to being consumers of, rather than stakeholders in, their own resources (Hampton 2003). Some provinces have chosen to restructure more than others. For example, Ontario (in 2002) and Alberta (in 1996) deregulated their electricity markets, and British Columbia is in the process of shifting new renewable generation (small hydro and wind) to the private sector (BC Hydro 2011, 2013; Calvert 2007; Province of British Columbia 2011).

There are two issues here. The first is the benefits in terms of cost and efficiency of private-sector generation and competition. The second is the definition of "green." What the empirical evidence in restructured markets suggests is that the consumers in these restructured systems face blackouts and higher prices (Beder 2003), large companies dominate and manipulate markets in their favour (Enron, most famously), and the small, green initiatives envisioned by environmental (and co-operative) advocates have a difficult if not impossible time getting on the grid (Walker 2008). This is particularly the case when restructuring is not aimed at increasing broad public participation.

Many environmental advocates support the restructuring of the electricity sector in the hope that new sources will be greener (Rifkin 2002; Scheer 2007) and will lead to a form of distributed generation (Walker 2008), thus breaking the concentration of power in centralized utilities (and, by extension, the nuclear industry). The participation of a range of non-traditional actors in the energy system (homeowners, co-operatives, local associations) encourages new innovation and competition and helps to develop resilience and self-reliance (Scheer 2007). Actors in the social economy and co-operative sector have also joined in support of distributed generation (CCA 2011a; FCPC 2013; Government of Canada 2012; Hoffman and High-Pippert 2009). For Newig and Fritsch (2009), this emphasis on localizing "expresses both a hope and an expectation that participatory processes will lead to improved compliance and implementation (measured against the agreed environmental goals) due to a more sound knowledge-base and an improved acceptance of decisions – in short: an enhanced *effectiveness* of the pursued policy."

While it is indeed true that new technologies open up the possibility of an alternative energy future, there is no reason that increasing the proportion of renewable sources will necessarily lead to distributed generation, a problem I return to in more detail in Chapter 4.[6] What is often lost in this discussion over greening Canadian electricity via private-sector development is that renewable hydroelectricity, while not without its critics, was highly developed by public (not private) utilities and accounts for more than 60 percent of Canadian generation by source.[7] Provinces today with the highest proportions of private ownership in electricity, Nova Scotia and Alberta among them, also have the heaviest reliance on carbon-based fuels. The reasons for this have much to do with historical trends and available fuels, though the contention that public electricity generation is somehow not "green" in a general sense is problematic. Public accountability and input into energy policies, as well as the ability to capture economic benefits for local development purposes, is vital for the development of sustainable energy futures (De Young and Princen 2012; Sathaye, Lucon, and Rahman 2011; Seyfang and Smith 2007).

Those arguing for an electricity market restructuring tend to maintain that governments are inefficient, cash-poor, slow to respond to market provisions or captured by private interests, or fail to provide consumer choice (Anderson 2009; Howe and Klassen 1996; IEA 2005). In this view, increased competition through privatization may drive prices down (an argument prevalent in the 1990s) and allow for a greater variety of generation sources (a more recent justification). There is significant debate over the effect that electricity market-led – as opposed to more participatory – restructuring will have on the development of renewable electricity.

There are also ingrained reasons electricity restructuring is taking place at this particularly historic juncture and manner, ones only tangentially related to environmentalism (Graefe 2006; Purcell 2008; Stanfield and Carroll 2009). First and foremost, ideological commitments of elected officials have in some cases directly legislated private-sector involvement in the electricity industry (e.g., in British Columbia, Ontario, and Nova Scotia). These moves came after years of the private actors being shut out of this increasingly profitable sector in Canada. Private firms have lobbied heavily to create and then access power markets. Consequently, attempts to green electricity generation that lack a broader understanding of political economy and public accountability risk a political naïveté that ultimately undermines progress toward deep sustainability.

Co-operative Electricity: Toward Empowering Power?

Re-embedding enterprises locally is one way to reconnect environmental and social feedback loops to democratic decision making. Doing so may be important in building resilience and empowering communities to address the complex challenges facing them in coming years. The electricity sector, though, has been incorporating more private actors, delocalizing, and generally failing in the development of greener alternatives.[8] Co-operatives, at least ideally, address many of the failings associated with conventional socioeconomic systems (CCA 2011b; FCPC 2013; Government of Canada 2012; Wright 2010). They are not as divorced from the real needs of Canadian communities and are, on the whole, organizationally more democratic. Indeed, co-operatives historically arose as local responses to the socioeconomic dislocations caused by the Industrial Revolution (Fairbairn 1990; Fairbairn and Russell 2004, 2014). This organizational alternative is not without its own challenges, however, as the ideal co-operative and co-operatives in practice often diverge.

The co-operative movement was one of the world's first social movements and is resurgent in many countries around the world (Curl 2010). The United Nations declared 2012 the International Year of Cooperatives. The movement is also far larger than most Canadians might think. Worldwide, over a billion people are members of co-operatives and, according to the UN, over half the planet's population is served significantly in some way by co-operatives (ICA 2015; MacPherson 2008, 640). Canada is no exception. As of 2007, one in four Canadians is a member of at least one co-operative (Co-operatives Secretariat 2010a). Co-operatives have played and continue to play important roles in community development and service provision across this country, despite their forming a largely forgotten chapter of Canadian economic education, conspicuously absent in business and economic texts (Kalmi 2007; Restakis 2010; Schugurensky and McCollum 2010). This oversight is significant, since these organizations not only make contributions to the material welfare of Canadians but also provide an institutional alternative rooted in norms that challenge neoliberal orthodoxy. In spite of their relevance, their influence on policy debates varies across the country and in different time periods. For example, federal budget changes in 2012–13 (during the International Year of Cooperatives, ironically) cut staff in the long-standing Rural Co-operatives Secretariat by 90 percent and ended funding to the successful long-standing Cooperative Development Initiative program (Government of Canada 2012).

Co-operatives – and the social economy more broadly – represent a pragmatic response to the economic and social challenges that both globalization and privatization have created. Co-operatives may make a significant contribution to the renewal of positive and active citizenship locally, nationally, and internationally (Lloyd 2007; Uluorta 2008). As an institutional form, the distinctiveness of co-operatives derives from an ownership structure of local actors based on community membership (stakeholders) rather than on financial capital (shareholders) (Quarter 1992). Since co-operatives are responsible directly to stakeholders, they may engender more environmentally sound and locally responsive practices (through local information transfer and social capital networks), empower underdeveloped areas (by pooling local resources), encourage entrepreneurial growth, and institutionalize an alternative economic rationality that explicitly links social and environmental needs to economic processes (Gertler 2001). In addition, co-operatives help address the principal-agent problem insofar as the users of a good or service also become the owners and sellers, resulting in a strengthened corporate framework to help avoid corruption and usury (Canada 2006; Mayo 2011; Neamtan and Downing 2005; Restakis 2010).

Quarter (1992) argues that associations and networks based on the norm of "people before profit" represent a key strength of this alternative system. Other writers on the subject support this position (Fairbairn and Russell 2004; Laville, Levesque, and Mendell 2007; Restakis 2010). The move to define co-ops as part of a broader global justice master-frame incorporating fair trade, local development, global institutional reform, and cultural exchange is an important one. Co-operatives engaging with these areas are part of what Vieta (2010) has called the "new co-operativism" based on solidarity and social justice, rather than on the narrower business models of many Canadian co-operatives. In this new kind of social and economic system, profit is but one of many goals, and participation, inclusion, and local development are paramount (McMurtry 2010).[9]

This connection between the spatial, the social, and the environmental has great appeal for bridging the often intransigent and thorny problems that transcend disciplinary boundaries of sustainability study and practice. It is here where the mutually constitutive and reflexive interactions between humans and their environment really hit the ground. In fact, recent research supports the argument that the types of interpersonal connections likely to occur on a local scale are, in fact, sustainability enhancing. For example, the Renewable Energy Consumption through the Community Knowledge Networks research group at Keele University in the United Kingdom has found that people are much more likely to change their behaviour based on information from friends, family, and local groups (including smaller companies), than they are if that information is provided by central government agencies or large companies. Face-to-face contact in particular helped people make informed decisions, as opposed to feeling overwhelmed or helpless in the face of complex and sometimes competing messages (Catney, Dobson, Hall et al. 2013).

Even given the potential of the co-operative ideal, serious questions remain about the role of the co-operative alternative within a broader neoliberal system of governance. First, co-ops have traditionally placed themselves (and been placed) somewhere *between* public and private sectors (Fairbairn 1990). Legally, they are private actors anchored by normative values of self-help and entrepreneurialism. Co-operatives are uniquely placed as locally owned businesses, to act as supportive alternative service providers for basic housing, health, and food needs that ameliorate the worst effects of state rollbacks in social services (Restakis and Lindquist 2001). This has the contradictory double effect of legitimating a discourse that private actors can handle these many important tasks while at the same time demonstrating that for-profit private actors abandon critical niches. Of course, despite co-operatives fitting within this frame, most are far from equal participants in newly opened markets. A significant tension thus exists between the co-operative ideal of a networked economic sector based on self-help and the more hierarchical organization of an interventionist welfare state. These two approaches to organizing society are by no means mutually exclusive, but the redistributive actions of an interventionist state sometimes stand at odds with a framework in which local resources contribute solely to local development.[10] This has created tension and debate over the political goals of the movement and over the relationship of co-operatives with the state and public policy (Amin, Angus, and Hudson 2002; Fairbairn and Russell 2004; Graefe 2006).

The co-operative sector in Canada today also lacks overtly political affiliations even though, in other countries and in earlier times – as with the Co-operative Commonwealth Federation (CCF) in Canada from 1932 to 1961 – the movement led to the formation of political parties (Laycock 1990; McMurtry 2004). In response to this passive role in broader political debates and processes, some have argued for more attention to be paid to how deeply public funding, regulatory structures, and policy affect co-operative and social economy organizations (LeBlanc 2006; Vaillancourt 2008). Finally, the ability to maintain a commitment to a meaningful level of democracy and broader social-movement awareness co-ops long term is questioned, as is the ability of these institutions to transcend the relatively marginal role they currently occupy in our economy (Fontan and Shragge 2000; McMurtry 2010).

Community Power
Co-operatives form one part of a broader community power movement in Canada that also includes First Nations, small business, and nonprofit development. "Community power," "community energy," and "community renewables" are terms used to variously describe institutional structures that include local input or control (Bolinger 2005; Devine-Wright 2011; ENVINT Consulting and Ontario Sustainable Energy Association 2008; Walker 2008; Walker et al. 2007). Yet, these terms are somewhat nebulous, as they can refer to a wide range of actors, ownership types, and forms of project participation. At the broadest end of the spectrum is the concept of community energy, which involves local collective action to generate or produce, distribute, and manage the energy resources of a community. This may include, but is certainly not limited to, the development of local energy

plans for reducing electricity or fuel consumption, municipal combined heat and power or district heating, the installation of solar or wind projects for either self-sufficiency or distribution and sale to à national grid, or non-profit or co-operative enterprises designed to provide energy-efficiency solutions such as insulation or consulting.

"Community power" – including community wind, community solar, and other derivations – refers to electricity-sector projects that involve local actors either in the design or operation of the project. It can also mean projects that are designed with community benefit in mind, but not necessarily deep engagement with the project by communities. Walker and Devine-Wright (2008) distinguish between various community power projects along two axes: whether the *outcome* is local and collective or distant and private, and whether the *process* is open and participatory or closed and institutional. A strong community project would be participatory as well as collective, whereas a utility wind farm would generally be characterized as more closed and private. A continuum of degrees of ownership is clearly possible, and regulatory requirements can open up planning and consenting processes to the public in important ways.

Typically, the community actors involved include Aboriginal communities; worker, consumer, or investment co-operatives; municipalities; non-profit societies; and farmers, as well as for-profit corporations made up of "local" residents of towns, districts, and sometimes even provinces. From this list, a few important differences stand out. First, "community" projects may not in fact involve or even benefit the majority of the local community. Benefits may be captured by a small number of local project investors or landowners. Second, "community" actors straddle the public-private divide, which involves different requirements for transparency, as well as obligations to the wider population of a project area. Still, we can identify broad differences in structure and interests between the various types of community power actors.

Municipalities and **First Nations** actors have clearly set-out obligations to their respective communities, as outlined in both provincial and federal legislation. Of the range of community power actors, these fit most easily on the public side of the spectrum, with clear and established relationships with other levels of government and an established institutional framework for projects. Project revenues and benefits are set to flow back into the larger organization for broader development and environmental sustainability purposes.

Private for-profit community projects can take numerous forms but generally consist of groups of local landowners or farmers forming corporations or partnerships to develop new renewable generation either for their own use (to reduce power bills) or to sell to the provincial grid for profit. **Nonprofit** projects are, as their name suggests, distinguishable from the others in that the key aim involves increasing uptake and development of new renewables rather than investor gain or other institutional goals (as with municipalities and First Nations, though their interests may overlap).

Co-operatives sit in a complex and overlapping space between the social goals of nonprofits and the profit motivations of corporate actors. Co-operatives' goals are driven by membership, rather than by either investors or the broader public, and embody the seven key co-operative principles, including one member, one vote; democratic control; and concern for community. The members consist of product or service consumers, producers, workers, stakeholders, or a combination thereof. Typically, renewable energy co-operatives are differentiated from more traditional co-operative forms in that members are more likely to play educative and investment roles than direct consumer or producer roles, and in that greater levels of financial investment and the capacity to withstand long lead-up periods is required. One can set up a renewable energy co-operative to be either a for-profit or a nonprofit entity. In the former, the project can recirculate financial returns from investments to members; in the latter, the social goals of the co-operative play a greater role, with returns being retained and redistributed by the co-operative for other purposes and projects. Another difference is that in for-profit co-operatives, investment shares and dividends (rather than bonds and interest) play a key role in financing.

This book focuses on the role of co-operatives in the electricity sector, rather than in the energy sector more broadly. This is in part because of the need to narrow the field of study given the plethora of co-operative models and actors with a general connection to "energy." This does not signify that the activities of co-operatives in heating, energy efficiency, fuel transport, or production – i.e., the broader energy sector – are less dynamic or relevant to the challenges of the triple crisis going forward. Indeed, as Chapter 6 illustrates, Canadian co-operatives are active in many areas, from oil refining and biomass generation to natural gas transportation and public sustainability education. In Denmark, wind turbine co-operatives emerged alongside the development of combined heat and power and district heating as part of a larger move to improve economic and environmental outcomes.

These are examples not only of diverse applications but of co-operative evolution, partnership, and "scale-up."

Denmark and Germany

The contribution of community ownership to new renewables development has been most striking in Europe. In 2013, "more than 3 million EU households produced their own electricity using solar PV (photovoltaics), and, by early 2014, 16% of Germany's businesses were electricity self-sufficient, up 50% from a year earlier" (REN21 2014, 80). Public policies in Germany provided strong incentives for new renewables development, supported by changes to co-operative legislation in the country in 2006. Germany is also in the process of phasing out its incumbent nuclear generation and has set a target of 100 percent renewable power by 2050. There were over 931 German energy co-operatives registered by December 2013 – more than 500 of these since 2010; nearly 90 percent of them are involved in developing renewable electricity (Debor 2014).

In Denmark, a country with the highest concentration of wind power in the world – 33 percent in 2012, set to increase to 50 percent by 2020 – co-operatives and farmer associations established the majority of the turbines (see Table 1.1). Wind power was not new in Denmark; a long history of windmill development together with the oil crisis, the Chernobyl nuclear disaster, and a willingness to experiment set the country on a unique path. In 2009, of Denmark's 5,200 wind turbines, 2,000 were owned by more than a hundred local associations. Electricity distribution was organized around locally owned (co-operative or municipal) organizations that are amalgamated at the transmission level into ten regional networks (Danish Energy Association 2009, 19; Stenkjaer 2008).

The obvious question here is *why* the Danish case is unique, and what lessons might this hold for Canadians (and others). First, many point to the important political and environmental debates that took place in the 1970s and 1980s in Denmark over the future role of nuclear power. Ultimately, a coalition of antinuclear, left, and green groups succeeded in making the case for a rejection of nuclear and the pursuit of other, more distributed technologies (Cumbers 2012; Danish Energy Association 2009; IEA 2012a; Lauersen 2008). Public-policy choices, including state tax incentives, supported a switch from fossil fuels following the energy crisis in the 1970s.[11] These policies focused on diversification of sources as well as on energy efficiency, dramatically reshaping the energy system from one that was centralized and fossil fuel reliant to one where decentralization,

TABLE 1.1

Comparison of international wind-generation ownership structures

Jurisdiction	Farmer %	Co-op %	Corporate %
Denmark	64	24	12
Netherlands	60	5	35
Germany	10	40	50
Minnesota	0	31	69
Great Britain	1	1	98
Ontario	0	<1	99
Spain	0	0	100

Source: Adapted from Gipe 2010.

local ownership, efficiency, and renewables play a significant role. Priority access for new actors formed an important piece of this policy puzzle, as did consumer price sensitivity. For example, Danish district heating policies specify that heat must be sold at cost (i.e., without a profit), and the system is run by community-municipality partnerships.

District heating together with combined heat and power (CHP or cogeneration) plays an important role in increasing the efficiencies of energy resource use in Denmark. The heat emitted from electricity generation in cogeneration is captured and used, rather than wasted. This heat can be used by the generation facility for its own purposes or integrated within a larger area as part of a "district heating" system, where centrally produced heat (geothermal and solar, as well as thermal power generation) is circulated through a local area, to be used for space and water heating. This system typically results in significantly reduced greenhouse gas emissions, cost savings, and resource consumption when compared with business-as-usual scenarios. In Denmark, co-operatives developed and operated district heating systems as well as wind-generation turbines and other electricity entities. According to a report from the International Energy Agency, "large power plants were again organized as co-operatives, with electricity distributors as owners. This form of organization, without a traditional profit motive, offered little resistance to government intervention in the sectors for electricity and heat" (Lauersen 2008, 1). According to Kerr, also at the IEA,

> The majority of the CHP plants serving the DH [district heating] networks are owned by local authorities and co-operatives, fuelled by natural gas. With so many individual households dependent on district heat, heavy

regulation of heat prices ensures that customer interests are protected. For example, the Heat Supply Law stipulates that DH schemes must operate on a non-profit basis, and heat and electricity prices must be cost-reflective. This fits well with the cooperative ownership of most DH schemes. (Kerr 2008, 4)

So, as with the development of wind power in the country, public policies used local ownership as a tool to increase both the provision of lower-cost energy and the uptake and support for new technologies.

The latest figures from the Danish Energy Agency (from 2013) illustrate the important role renewable energy sources and efficiency technologies play. Renewables (solar, wind, hydro, and biomass) make up nearly 41 percent of total electricity use and 23 percent of total energy use. Moreover, wind turbines account for 29 percent of total electricity generation capacity, and CHP production makes up 76 percent of total district heating (Danish Energy Agency 2013). Denmark is now a world leader in renewable-energy policy design and in energy co-operative development. It has the highest share of CHP and district heating systems in the world, which, together with the discovery of North Sea gas, allowed the country to become energy self-sufficient in 1997. Because of these and a range of other changes, Danish GHG emissions per capita for 2012 sat just at 9 tonnes of CO_2 equivalent, down from 13.4 in 1990 and half that of Canada's (OECD 2015, 18).

The evolution of co-operatives in Denmark is illustrative of how electricity co-operatives can operate beyond the binary of electricity generators and efficiency advocates. It illustrates the importance of understanding the varied co-operative/community power form. For example, co-operatives in Denmark are divided between large co-operative CHP plants and smaller co-operative wind companies. In the former category, six of the ten largest (coal-fired) power generators in the country, including Syd Energi and SEAS-NVE, are actually co-operatives whose members are electricity distributors (Kerr 2008; Lauersen 2008). In the area of wind generation, the community power co-operatives are technically wind-power stations or general partnerships. According to the former chairman of the Danish Wind Turbine Owners' Association, "for legal reasons [the co-ops] were forced to make formal partnerships due to the fact that, in Denmark, the interest on the loan for the wind turbine is tax deductible from the private income of the individuals in a partnership, not in a co-operative. Danish Wind Power Stations tried for years to have the law changed on this point, but did not succeed" (Tranæs, n.d.).

Community ownership of new turbines has decreased since 1995 along-side increases in turbine sizes. Recent research on the Danish case illustrates how a shift in public-policy supports for community power in the late 1990s has led to social friction over turbine development and has stalled new projects through the mid-2000s (Cumbers 2012). This trend away from local ownership has been identified as problematic and attributed to a shift toward more free-market electricity policy in the country, as well as to the maturation of the wind industry (Larsen 2005; Lauersen 2008; Manczyk and Leach n.d.; Meyer 2007; Möller 2010). In a report for the Danish Energy Agency, Jensen and Jacobsen (2009, 8) point out that "the progression toward fewer joint-owned and relatively large turbines has made it difficult to maintain support for new windpower projects." Indeed, the Danish Promotion of Renewable Energy Act of 2008 aims to address these issues by mandating that wind developers offer at least 20 percent of the project for sale to local populations. It also set up funding for municipal improvements around wind parks and a local ownership start-up fund of DKK 10 million (approximately 1.9 million Canadian dollars) for preliminary studies (Jensen and Jacobsen 2009).

Danish and German experiences with community and co-operative power have led to significant networking and policy learning about community renewables development between Canadians and their European counterparts. One clear lesson from these developments is that public-policy supports formed a crucial element of these community and co-operative systems[12] (Bolinger 2005; Walker 2008). Another is that just as supportive policies can emerge, they can just as quickly be reversed if the political winds change. Beyond that, there are a significant range of structural factors related to resource endowments, industrial structure, political cultures, and policy regimes that are likely to affect the success of a Danish model in other jurisdictions (Bolinger 2005).

Canadian Electricity Co-operatives

Electricity co-operative potential is reliant on targeted policy changes that support community power. Within the range of policy options, organizational structures, and actor goals, a wide range of tensions emerge. At the heart of the community and co-operative power movement is the contention that local involvement in energy projects is both necessary and desirable. There is a large and growing literature on the contribution that direct ownership of resources through community and co-operative power makes to communities (Bolinger 2005; ENVINT Consulting and Ontario Sustainable Energy

Association 2008; Gipe 2007; ILO 2013; Jacobsson and Lauber 2006; Warren and McFadyen 2010). In short, there are five core arguments for social ownership and control of resources: social economy energy provision (1) combats "not in my back yard" attitudes (NIMBYism) by giving locals a stake in the project, (2) helps educate communities about their resources, (3) spurs local development and job creation, (4) keeps profits in communities and builds local capital (financial and human), and (5) provides legitimacy to renewable energy projects.

Electricity co-operatives are a resurgent development in most Canadian provinces. Hundreds of rural electricity co-operatives formed between 1940 and 1960 in Alberta and Quebec. But this provincially concentrated picture has changed over the past thirty years (accelerating in the past ten) as their development shifts east, to Ontario, Quebec, and the Maritimes. Today, electricity co-operatives exist in every Canadian province. They are developing in both urban and rural areas, and are engaged in generation and distribution, as well as in education and retail of new renewable electricity (e.g., solar and wind generation). They are increasingly networking with other renewable electricity players to lobby provincial governments as part of efforts by the community power sector to achieve market support (primarily feed-in tariffs, or FITs[13]) for locally based private renewable development (FCPC 2014, 2015; Lipp, Lapierre-Fortin, and McMurtry 2012).

Electricity co-operatives take numerous forms. One is the generation of power that is then transmitted through the grid and sold to public utilities or private retailers. In Canada, the vast majority of co-operatives working on developing generation have focused on wind and solar power. A second model exists in which co-operative members pool their assets to build (or buy) sections of the distribution system. These co-operatives are concentrated mainly in Alberta and are divided between ones that own *and* maintain the distribution system (self-operating distribution co-ops) and those that own the lines but contract out to other players in the power sector (e.g., Fortis and ATCO Electric) to manage the lines for them. Co-operatives in this sector can also be structured as consumer pools to buy bulk electricity – possible in deregulated retail markets such as Ontario and Alberta – for their members at a lower cost. As well, consumer electricity co-operatives can source products for their members in order to encourage such things as solar panels on housing and energy conservation. In the power sector, worker-owned co-operatives are rare, but possible. At this point in Canada, they are mostly sustainability consulting businesses, but there is a project in Quebec that generates power using biomass from wood waste. Finally, electricity co-operatives can be structured as nonprofit community associations. These focus on conducting educational campaigns for sustainable and renewable energy, and sometimes, as in the case of the Toronto Renewable Energy Co-operative, act as an incubator for generation co-operative project spinoffs.

Co-operative electricity-generation projects in Canada are just starting to take operational shape. Although some communities have been actively pursuing projects for almost ten years, a range of problems – from grid connection to policy supports to volunteer burnout – have resulted in relatively few projects actually being built. As with all players in the electricity sector, but perhaps more so than most, co-operatives are dependent on state choices; public-policy decisions significantly affect the strength of industrial competitors for co-operatives, as well as market prices and the very basic legislative and legal support for the co-operative form. For example, governments grant co-ops legal and tax status, provide subsidies for local economic or environmental projects, and grant access for electricity co-ops to the distribution grid. In fact, many electricity co-operative projects attempting to connect to the grid in Ontario's Orange Zone (an area where transmission has reached capacity) are stalled because of provincial agreements with nuclear power providers.

This picture may be set to change somewhat as jurisdictions across the country – Ontario, Quebec, New Brunswick, and Nova Scotia – are starting to support co-op developments. In 2011, twelve projects either owned or initiated by co-operatives were generating electricity across the country.[14] Many more are in the project-development phase: nearly 105 MW of installed capacity from wind-power generation co-operatives in New Brunswick, Ontario, and Quebec was awarded power purchase agreements in 2010 (total installed wind capacity in Canada, by comparison, was 8,517 MW in September 2014. According to the Federation of Community Power Co-ops in Ontario (FCPC 2015, 18–21), between 2012 and 2015 in Ontario, a further 108 MW went to co-operative generation projects in FIT rounds, and these organizations hold roughly $94 million in assets. Investments in these projects ranged from $1,000 to $150,000. More than 1,000 separate microFIT contracts in Ontario were also awarded to co-ops[15] (OPA 2012, 2013a, 2014; FCPC 2015). These projects, and many others like them, are explored in more detail in Chapters 6 and 7.

The benefits of co-operative electricity projects transcend material (financial and service provision) benefits. They play a symbolic role in shaping public perception of the possible. Community electricity projects can be used as demonstration projects and as educative tools to engage broader audiences. This value is often cited by participants and initiators of these projects as a driving goal (Ferrari personal interview, July 23, 2009; personal interview, July 23, 2009; FCPC 2015); it extends beyond monetary gain to the transformative role that projects càn play in shaping public opinions, experiences, and, through that, policy. Indeed, the interactive role between the constituencies created by community groups and policy change is well documented (Walker et al. 2007).

The relationship between co-operatives and the broader private sector is both important and problematic. A few key limitations affecting new electricity-generation co-operatives illustrate this. First, rarely are generation projects wholly owned by co-operatives. Ownership and control ranges from 100 percent co-operative, as in the Ottawa Renewable Energy Co-operative, to a minority share in a limited partnership, such as the one between Peace Energy Co-operative, Aeolis Wind Power, and AltaGas on the Bear Mountain Wind Park in British Columbia. A sliding scale thus exists, with a project solely owned by members at one end, and a project owned by a private- or public-sector entity at the other. Most projects are a combination falling somewhere in the middle. Private-sector partners are sometimes keen to work with community-based groups such as co-operatives because they help provide local legitimacy for a project and aid in getting through the environmental assessment and consultation stages. In an industry where years of feasibility studies and approvals are necessary, it can mean significant amounts of wasted time and money if local resistance leads to the cancellation of a project. At the same time, private-sector involvement raises the issue of co-optation of community projects and attendant concerns over the political role of community energy within the broader power sector.

Access to capital is a second challenge and one of the main drivers behind the partnership strategy of project development. It is an especially important issue for generation projects, since they are capital-intensive and require years of development and testing before the returns are realized. This means that a financing structure that recognizes the benefits of community-based enterprise is essential in Canada. Without this, community groups are often restricted to developing either very small (one turbine) projects or to partnering with larger developers (with reduced control and stake). In Germany, for example, farmer-owned wind projects were feasible because the government gave loan guarantees to farmers to develop their wind resource (Gipe 2007; Toke, Breukers, and Wolsink 2008). This gave banks the confidence to lend and the farmers access to much-needed capital without ceding control to nonlocal developers.

Further assessment of where and how significant contributions are being made by Canadian co-operatives is important in light of challenges and successes in other jurisdictions. What is clear is that the shape and success of these co-operatives is dependent on a wide range of political economy factors. These organizations are providing legitimacy, via community buy-in, to broader shifts toward electricity restructuring and thus play a role in shaping sectoral social, economic, and environmental impacts. This impact occurs whether the ultimate projects are successful or not, though certainly not to the same degree. Electricity-sector restructuring has enabled generation co-operatives, as independent power producers (IPPs), to sell community-based energy to the grid via standard contracts. These electricity co-ops have become possible only as provincial governments open up electricity markets to private actors and energy trading. This raises interesting questions of how co-op actors today view the shifting power in this sector, and the value of public ownership of utilities more generally. Co-ops in this area face not only the challenges of sustainability, visibility, and support but also competition with some of the most powerful corporations in the world. Whether and how they learn to overcome these challenges will provide an important test for any local electricity alternative. The paucity of data on electricity co-operatives, together with the fact that they straddle a range of research areas, makes for a pressing yet rewarding research challenge.

Overview of the Book

Electricity co-operatives in Canada are seeking to develop community-based electricity guided by principles of democratic decision making and local stakeholder – rather than shareholder – control. These co-operatives represent an alternative form of renewable electricity development. Each chapter of this book develops further the core propositions laid out in this introduction, namely that the restructuring of Canadian power sectors is taking place and that the policy choices made impact not only co-operative development and potential but also the safety and security of Canadians. This is, in part, because the green power movement is contributing to a shift away from public ownership of new generation. Electricity co-operatives are enabled by these developments. However, established energy lobbies in the fossil fuel and nuclear sectors retain significant policy influence, so that newer actors and sources face a steep uphill battle. So, despite their democratic and local appeal, co-operatives are significantly constrained by both internal and external factors in their ability to provide a significant power alternative.

The first two chapters set the structural and theoretical context for my analysis of electricity co-operative development. In Chapter 2, "Governing Sustainability: From Crisis to Empowerment," I develop a conceptual framework for understanding these electricity co-operatives that embeds their development in a political economy understanding of the often contradictory processes and forces of neoliberal governance. This framework is built with an interest in understanding the ideological and material processes that inform not only co-operative developments in the past but also the potential for these institutions going forward. Chapter 3, "Co-operatives in Canadian Political Economy," presents the argument for how, why, and where co-operatives may form a more democratic and empowering alternative to other forms of organization in Canada, and explores some of the contradictions and challenges accompanying this form of organization.

The next two chapters focus on developments driving change across provincial power sectors in Canada. Chapter 4, "International Forces for Power-Sector Restructuring," illustrates how ideologically driven policy choices have prompted a global trend toward restructuring of power sectors in countries around the world. That is, the pressures and changes taking place in Canada are part of global neoliberal processes, embedded in and facilitated by international and continental institutions like the World Bank, Organisation for Economic Co-operation and Development, and Federal Energy Regulatory Commission. Chapter 5, "Continental, Private, and Green(er)? Canadian Electricity Restructuring," links empirical developments toward new private, green, and community-based power in Canadian electricity to neoliberal governance. In it, I investigate provincial variation in electricity-sector ownership and generation sources with a view to situating co-operative development solidly in the material basis of a given electricity regime. I argue that public policy, not technological or financial necessity, prompted power-sector reforms across Canada. These reforms are accompanied by challenges for the rapid energy-sector transformations required to mitigate ever-increasing GHG emissions. These reforms also bring challenges for both electricity ratepayers and communities more generally.

The second half of the book moves from the political economy context to examining co-operative development in provincial electricity sectors. Chapter 6, "Electricity Co-operatives: The Power of Public Policy," charts the development of electricity co-operatives across Canada from the 1940s up to 2013, and highlights similarities, differences, and the diverse contributions these organizations have made through periods of electricity-sector development. Overview data is presented on the total population, geographic distribution, and diversity of electricity co-operatives existing in the country. Chapter 7, "Off the Ground and on the Grid: New Electricity Co-operative Development," examines in more depth the promises and pitfalls of recent electricity co-operatives, particularly those participating in renewable electricity generation. These include benefits of local economic development and sustainability education, as well as the challenging issues of financing, grid access, and community capacity.

In Chapter 8, "Co-operative Networks and the Politics of Community Power," I examine the participation and role of co-operatives within policy networks and tensions within the community power movement. The organizational diversity of co-operatives – distribution, generation, consumer, and networking – leads to diverse roles in the electricity sector.

Finally, in the concluding chapter, "Empowering Electricity," I return to the challenge of developing more democratic, green, and local electricity systems in Canada. Although certainly contributing to community development and control in specific instances, they are at present far from a significant challenge to the broader involvement of for-profit private actors in the electricity sector. This challenge does not preclude future promise though, and this final chapter explores how electricity co-operatives may "scale up" to play an important role in sustainable electricity futures.

2

Governing Sustainability

From Crisis to Empowerment

Empowered participatory governance provides a useful theoretical perspective through which to analyze recent co-operative developments. It addresses the role that economic structures and institutions play in transforming human impacts and behaviours, harnessing a diverse and interdisciplinary set of perspectives from political economy, policy studies, political science, and environmental studies traditions (Johnston, Gismondi, and Goodman 2006a; Kumhof and Ranciere 2010; Williams 2010). Governance is put front and centre conceptually because it problematizes the different geographic levels (local/national/global), actors (public/private/network), *and* ideologies (neoliberal/ socialist/communitarian) at play structuring socioeconomic relations at a given point in time. Although this lacks parsimony, analysis of each of these – levels, actors, and ideologies – is required in an analysis of the empirical contribution electricity co-operatives can make to sustainable futures.

The concept of sustainability entails a capacity for the endurance of human social and life-support systems. The Brundtland Commission definition of "sustainable development" as "meeting the needs of the present without compromising the ability of future generations to meet their own needs" sets a baseline for understanding sustainability as an issue of needs (versus wants) and of fair distribution between different groups in space and time (Brundtland 1987). Transitions to sustainable governance thus require maintaining stocks of natural capital, a degree of intergenerational equity, and a restructuring of the relationship between the environment, economy, and society (Barry 2012; Davies 2009; Lehtonen 2004). It is widely acknowledged, however, that the latter "three pillars" – environment, economy, and society – approach to sustainability is problematic insofar as the pillars are misconceptualized as independent rather than embedded and the "social" aspects are routinely undertheorized and underemphasized (Blay-Palmer 2011; Cuthill 2010; Rees 2010).

Widely used conceptual silos – politics and economics, local and global, and social and environmental – serve to limit our capacity to respond to current challenges, and so in this book I draw on theories that emphasize holism, historicism, and multiple sites for collective action. Exploration of different modes and practices of governance opens up new sites of inquiry into actors and processes that fail to neatly divide along the private/market versus public/state dichotomy. In reality, actors in many sectors guide stasis and change through networked governance, epistemic communities, and sometimes even outright regulatory capture. Hybrid actors situated between the conventional public and private spheres may play an important role in transitions toward new systems by demonstrating a wider possible range of economic organization, and by helping diffuse a set of norms around social and environmental obligations throughout a population.

One of the ways to gain analytical traction within this complexity is to focus on the genesis and impact of changing government policies. For example, policy shifts can support the emergence of strong new constituencies and new technologies, and alter the balance of power in a given subsystem (Howlett, Ramesh, and Perl 2003; Rotmans, Kemp, and van Asselt 2001; Seyfang and Smith 2007; Vaillancourt 2008). We need to understand not just the endpoints of a transition in terms of the number of organizations and share of new renewables but also the new processes and norms developed. To do this, this chapter examines some of the questions raised in Chapter 1: What kinds of institutions are likely to deepen both democracy and sustainability in our society? How can we evaluate these institutions, and what challenges might we expect to emerge in terms of credibility and efficacy?

I draw on a heterogeneous collection of theorists to answer these questions, ones whose perspectives interrogate the structures of political and economic power underpinning co-operative development. Some of these perspectives, described below, come from institutional theory (Ostrom 1990, 2002), governance (Jessop 1995; Rosenau 2003), and eco-political theory (Johnston, Gismondi, and Goodman 2006a); others from theories of crisis and transformation (Cox 1996; Harvey 2010; Wright 2010), capitalism (Polanyi 1944), and community economic development (Loxley 2007). What unites these theorists is a perspective of socioeconomic change that is historical and dynamic. Theories of change that are mechanistic, stressing inevitabilities and irreversible processes, are problematic, as they underestimate and misrepresent the role of human agency, contingency, and contextuality. An understanding of community-level institutions and change that is both theoretically informed and respectful of the real work being done on the ground, often by volunteers, is vital. The analytical approaches outlined in this chapter provide a framework not only for why institutional innovations like co-operatives are theoretically interesting but also for what particular aspects of their structure and activities may hold the most potential.

Neoliberal Governance

Rosenau (1995, 14) defines governance as that which "encompasses the activities of governments, but it also includes the many other channels through which 'commands' flow in the form of goals framed, directives issued, and policies pursued." Studies of governance, as with those of political economy, emerged out of the understanding that the exercise of power in society covers much more than what governments do, and that market-based actors play important roles in directing both stasis and change (Andersson and Ostrom 2008; Held and McGrew 2002; Jessop 2002; Steurer 2013). A focus on governance captures the institutional configurations at multiple levels beneath, within, between, and above states, whether public or private.

Understanding these complex pathways of power is essential when private-sector actors are increasingly central to public-policy making and permeate most aspects of socioeconomic life. In practice, there is no governance without government; the private sector is, even in a neoliberal era, reliant on public-sector power to create new spaces for private accumulation, and to financially guarantee investment in new and risky ventures such as renewable electricity development. Public-policy choices underpin this neoliberal shift, and identifying the specific actors and forces behind them is important (Harvey 2005; McBride 2005). These actors may come from the formal public and formal private sectors, as well as from sectors that sit between these poles, like co-operatives and social economy organizations. Ownership of assets and the degree of collective control over them matters for how governance takes place: how permeable, how participatory, and how flexible (Baker 2014; Peris, Acebillo-Baque, and Calabuig 2011). Broadening the theoretical lens in this way to include the complex of actors and mechanisms where power resides facilitates examination of the real, rather than ideal, channels through which crises are created and transformations can occur.

The dominance of neoliberalism – defined here as an ideology of socio-economic governance premised on a reliance on market allocation of resources, the rollback of state expenditure on public services, privatization, and deregulation – in the Canadian political economy has underwritten the policy shifts that enable electricity co-operative development. Peck and Tickell (2002, 380), for example, suggest that "neoliberalism provides the 'operating framework' or 'ideological software' for competitive globalization, inspiring and imposing far-reaching programs of state restructuring and rescaling across a wide range of national and local contexts." This software includes a powerful and now ubiquitous rhetoric that constructs a narrative of governance without government, and stresses the virtues of private competition, individual choice, and freedom from regulation, taxes, and responsibility to a collective public good. In the electricity sector, this "operating framework" emphasizes the utility of privatizing public utilities (or parts of them), entrepreneurial innovation from private citizens, and the use of market incentives to spur private generation of new renewables where necessary.

The particular ideas underpinning neoliberalism shape governance in Canada today. Many scholars argue that there has been a marked shift in the ideas and institutional configurations (or paradigms) shaping Canada in the past fifty years (Clarkson 2002; McBride 2005; Panitch 2007). Governance has shifted power from public hands to private, and governments both provincial and federal have ceded authority to international bodies. This change in modes of governance, from Fordist to post-Fordist, and from Keynesian to neoliberal, is significant. Larner and Craig (2005, 4) argue that "a mode of governance is ... a set of rules, a set of knowledge and a structure of collaboration for day-to-day decision-making. It includes the social world that is part of these practices in terms of both the subjectivities of the actors and the material objects that are produced."

Neoliberal modes of governance arise from ideological norms embedded in governing institutions as well as from the concrete structural, material, and legal relationships of power between various actors. These relationships are embodied in specific institutional, policy, and legal changes (Larner and

Craig 2005; Peck and Tickell 2002). Peck and Tickell (2002, 37) developed a useful account of how neoliberal ideology and governance translates into policy initiatives. They distinguish between "rollout" and "rollback" neoliberalism, though the two are interrelated. Rollout neoliberalism involves, for example, accumulation by dispossession, commodification, and bringing new goods into the market economy. Rollback neoliberalism, on the other hand, involves scaling back and constraining state agencies through policies of privatization, deregulation, and a general discrediting of Keynesian institutions.

Cross-sector (or public-private) partnerships represent a further policy manifestation of neoliberal governance. Forsyth elaborates on how these use the language of participation, democracy, and empowerment to devolve (and often privatize) formerly state functions:

> Accordingly, cross-sector partnerships (CSPs) have become something of a panacea for some analysts – often those proposing neoliberal, or New Public Management approaches to public policy – because they attempt to empower individuals and businesses within public policy, while also diminishing the reliance on states. Indeed, CSPs form part of a growing trend toward a more deliberative and devolved form of governance using concepts such as "public policy partnerships," the "mutual state," or "network" or "hybrid" governance. These approaches, in principle, aim to harness civil society more effectively within public policy by increasing public debate, and passing greater responsibility for certain public services to the local level. Proponents claim doing this will increase the speed and accountability of local public service provision, and decrease costs by reducing the need for a centralized state. (Forsyth 2010, 683)

Whether in rollout or rollback form, maintaining and extending neoliberal governance requires the reinforcement of capitalist norms through social and cultural projects emphasizing individualism, competition, and marketization.

Electricity co-operatives, and co-operatives more generally, form a growing part of rollback neoliberalism. Where Crown corporations have their service mandates scaled back or delisted (as with wind generation in British Columbia), co-operatives can participate with other private actors in newly opened markets. In other areas, like health care and social housing, co-operatives have also stepped in to fill roles in public-private partnerships and alternative service provision (Graefe 2006). Co-operatives, nonprofits, and other social economy institutions are well suited to play these gap-filling roles, requiring less profit, with generally local and democratic organizational structures, and a tradition of volunteerism and service. Importantly, though, institutions like co-operatives also play an important role in mediating, translating, and sometimes challenging socioeconomic norms, dominant structures, and processes. Although in one sense they are responding to the rollback policies initiated in other areas of a given society by policy makers and private-sector actors, they are also active participants in accepting or reshaping these processes.

Sustainability, Capitalism, and Crisis

Radical critiques of contemporary neoliberal governance are rooted in a holistic understanding of the social and environmental foundations of economic systems. Capitalism, at the most basic level, places private ownership and markets, rather than a broad range of interests and polities, in the control room. This can lead to the kinds of destabilized systems and socioecological crises that have emerged today. Three central points raised by political economists are relevant to our understanding of the roots of the triple crisis described in Chapter 1: (1) economic crises are endemic to capitalist economies; (2) both social and environmental degradation are a result of commodification and exploitation of human and natural resources; and (3) crises create opportunities, and from within these contradictions new and potentially transformative forms and modes of socioeconomic activity may arise.

First, economic crises are endogenous to the capitalist mode of production and accumulation. Harvey, for example, compares many of the theories of crisis that were raised after the global financial crisis of 2008. Mainstream accounts ranged from Alan Greenspan's contention that people are essentially greedy individuals to the blaming of regulatory institutions for not being "Keynesian enough" or, conversely, "regulating too much" (Harvey 2010; Patel 2009). Harvey, like Marx before, challenges these mainstream accounts and locates the cause of current crises in the circuits and processes of (particularly neoliberal) capitalism itself. The constant need for growth, for more profits at faster and faster rates, pushes capital to continue expanding ad infinitum. This requires creating and opening new markets and spaces for accumulation through privatization, financialization, and accumulation by dispossession. These processes, in turn, lead to destabilization and volatility and, ultimately, crashes and crises. The Enron scandal – wherein new deregulated markets led to blackouts, soaring electricity prices, and (illegal) predatory practices by energy traders – represents but one example of this.

Second, social and environmental crises are tied to the structures and processes of neoliberal governance (Albert 2003; Bowles and Gintis 1986; Kovel 2007). Polanyi's (1944) account of the development of the idealized self-regulating market in *The Great Transformation* is instructive, as he, like Marx, focuses analysis on the relationship between the economic and social systems. For both, albeit in different ways, the market is intrinsically a structure of power. Polanyi illustrates how social dislocation and ecological destruction – processes coming to a head today – are the result of developing an economic system that is guided by the myth of a self-regulating market. He argues that by turning nature into a "fictitious commodity," the now-mainstream market capitalist system erodes the very basis for human survival. Thus, economic democracy is essential and is achieved by asserting social control over the economy and society at multiple levels and across issue areas. Marx and Polanyi also recognize the key role that nature plays in human development, and how it is threatened by exploitative systems of production and consumption (Burkett 2006).[1] Whereas most accounts focus on the social aspects of exploitation in alienating humans from the fruits of their labour, their political economy analyses continue to be relevant today to how we conceive of the root causes of social disempowerment and ecological degradation (Fitzpatrick 2002; Foster 2002; Williams 2010).

Given these important interconnections between economic and ecological crises, it is problematic that environmental and radical political economy literatures have talked past each other until quite recently (Brecher, Costello, and Smith 2008; Burkett 2006; Faber 2008). In particular, literature from the deep ecology perspective takes the view that nature is not a resource to exploit but is instead the fundamental basis for all human life, with inherent value of its own (Luke 2002; Næss 1973). This is a valuable contribution to contemporary debates over the triple crisis, as it situates human society within biophysical limits. Eco-centrism in the deep ecology literature, however, fails to adequately address issues of human justice and distribution such as employment, class, and north-south exploitation as they relate to environmental policy. On the other hand, traditional left analyses too often fail to accept or develop an understanding of ecological limits to growth, even though the seeds of this exist in the work of both Marx and Polanyi (Foster 2002, 2009). The result of this gap has been the dominance of free-market green politics (Bakker 2010; Driesen 2009; Heynen et al. 2007). With the rise of the environmental justice movement, this dominance is being challenged, and new sites of debate are emerging over the appropriate values and institutions of sustainability. Two potentially fruitful lines of inquiry that draw both together are eco-localism and eco-socialism (Curtis 2003; Sandberg and Sandsberg 2010; Vanderheiden 2008; Williams 2010). The former connects environmental outcomes to place-based governance; the latter links common ownership with collective management of the commons.

Third, crises lead to ruptures in traditional systems, allowing for rapid systemic change wherein new forms emerge from the contradictions and challenges of old systems (Homer-Dixon 2007). Cox (1996) argues that it is the crises and contradictions, which open up ruptures between political powers and economic systems, that then allows for the seeds of a new system to take root. Whereas neoclassical scholars treat crises as exogenous to the system, in the neo-Marxist literature, for example, theorists in the Social Structure of Accumulation (SSA) (or Régulation) school highlight both the centrality of crisis in capitalism and that the roots of its transformation lie within current systems (Boyer and Saillard 2002; Jessop 1995). Institutional variation within capitalism is thus important in producing different sorts of arrangements, contradictions, norms, and opportunities for substantive change (Barry 2012; Munck 2002; Williams 2010).

This understanding of the genesis of transformation echoes Polanyi's (1944) insight in what he called the "double movement": the reassertion of social control over market forces via protective legislation or restrictive associations (such as co-operatives). For Polanyi, asserting social control over the market society at multiple levels and across issue areas is a natural response to the expansion of markets, particularly when they threaten society and lead to crisis. In order to understand the creation of new forms of socioeconomic control, one need only look to areas where the market is displacing earlier forms of organization: provincial electricity sectors in Canada, for example.

Ambiguities regarding potential agents and precise drivers of change present a challenge for those advocating a deep transformation of society. Theorists have variously located this agent with unions and organized labour (Gamble et al. 2007), community associations (Scholte 2003), collectives (McMurtry 2004), and environmental and peace movements (Della Porta and Tarrow 2004). Both material and ideational developments play a role in shaping social change. Gramsci (1971), for example, highlights the crucial role of ideational factors – in particular, the role that consent plays in upholding relationships of hegemony and force – in provoking or retarding radical social change. Thus, it is not only the material changes in the structure of ownership and production that guide transformation but also the norms and ideas legitimating material structures.

Building Alternatives: Governance, Institutions, and Countervailing Power

A rich and growing literature seeks to reconnect economic governance to social and community roots and to redefine and revitalize the theories and practices of democracy (Adkin 2009; Harcourt and Wood 2004; Lambert 2007; McMurtry 2004, 2010). These represent an alternative to what Cox (1996, 303) calls the "limited democracy" of neoliberal governance. He argues that occasional elections and consumer choice do not equal democracy in a meaningful sense. Substantive democracy, therefore, requires a relinking of economic policy to social and political control via the formation of institutions and norms of *economic* democracy. It also requires deliberative and participatory policy mechanisms that are inclusive, informed, and uncoerced so that citizens have a more legitimate say in their own governance (Bäckstrand 2010; Dryzek 2002; Johnson 2008). Finally, transitioning to substantive democracy requires a move toward mutually reinforcing and multilevel participatory institutions or "empowered participatory governance" (Fung and Wright 2003).

Dominant rules and principles, together with the actors and structures, matter to the shape of future transitions (Andersson and Ostrom 2008; Forsyth 2010; Wilkinson and Hughes 2002). Co-operatives, and the social economy more generally, play an understudied role in socioeconomic transformations as locally and socially grounded institutions. The mesolevel of institutions transforms microlevel individual behaviour as well as macrolevel forces. This level of analysis where agency and structure meet is critical for an understanding of which specific actors may be useful in bringing about transformation. This perspective is often missed in the literature on crisis, which focuses more on diagnosing the structural problem than on understanding the agents of change. On the other hand, a focus on individuals and individual-level rationality – typical in economics and increasingly in political science – is also problematic. Values and structures underpinning particular organizations can have an effect down to the people who are members of them and up to the governing structures (Fung and Wright 2003).

Elinor Ostrom's Nobel Prize–winning work has been illuminating here. She argues that institutions, collective action, and rationality have been both understudied and misunderstood (Ostrom 1990, 2002). She questions the need for state control or private ownership and focuses on the role that participatory, local, and democratic institutions play in managing resources sustainably (1990). Furthermore, the structures and forms of co-operation within which collective action takes place are critical for informed political analysis. These structures include locally based management systems for resources, such as co-operatives. Ostrom, as a critic of simplistic economic and institutional thinking, argues that co-operation and property rights regimes are thus a critical site of inquiry. This is because social networks, communication, and trust are crucial for overcoming collective-action problems (1990). Co-operatives represent an institutional variation within the current capitalist system that carries norms and practices distinct from more dominant institutional forms in conventional public or private sectors (Amin 2009; Smith 2009).

Localism and Sustainability

Locally rooted co-operative institutions may thus play an important role in facilitating this kind of participatory democracy: their typically smaller scale often (but not always) allows for more deliberative space, room for development of capacities, and the inclusion of marginalized groups. The theory and practice of eco-localism – the view that environmental sustainability is best advanced by local self-reliant economic communities – also sets a foundation for interest in and support of electricity co-operative development (Curtis 2003; De Young and Princen 2012). Various elements of this thinking collectively emphasize the importance of local democracy, resource constraints, creativity, sufficiency, and heterogeneity. These potential theorized benefits of eco-localism range from technological innovation to enhanced civic participation, from lower-stress workplaces to healthier food and communities, and from reduced resource use to local resilience and adaptive capacity to disasters (Folke 2006; Hahnel 2007).

This eco-local perspective is supported by environmental advocates of Schumacher's (1973) *Small Is Beautiful,* which sets out a vision of a future of small-scale communities operating with appropriate technologies and within ecological limits. The founder of ecological economics, Herman Daly, emphasizes the importance of bioregionalism in his work as a step toward living within the carrying capacity of the earth (Daly 1996; Daly and Cobb 1989). Of particular importance to advocates of localized governance is the role of changing human behaviour via enhanced civic participation in resource allocation. For De Young and Princen (2012, 328), "localization does not accept the dominant assumption that increasing affluence is an unmitigated good, nor that lowering material wealth necessarily decreases well-being. In fact, affluence and well-being presumed to go with it can be the product of brutal working hours and workplace competition, inhumane levels of stress and uncertainty, and destructive interactions with natural systems." As Catney, MacGregor, et al. (2013, 1) argue, "a broad sweep of the ideological spectrum now advocates localism as a primary vehicle for, and even a goal of, new forms of social, political and economic organisation." This perspective stands in stark contrast to policy initiatives in most states that emphasize economic efficiency, growth, global trade, and homogeneity (as well as substitutability).

Unpacking the democratic and environmental theoretical commitments of this eco-local view is important for our conceptualization of both co-operative potential and empowered participatory governance. The first set of connections links democracy, participation, empowerment, and local institutions. Democratic theorists such as Pateman (1988) – following Aristotle and Rousseau – have held up participation as a crucial pillar of democracy, wherein citizens share in decision-making processes in a meaningful (rather than consultative) way – meaningful in the sense that the issue areas open to public control extend into social and economic policy, rather than just through infrequent elections. It is also meaningful in the sense of a deeper level of participation that includes debate and deliberation (Johnson 2009; Wright 2010). Pateman (1988) argues that for a healthy democracy, people need to learn the skills of self-government through, for example, local processes like participatory budgeting. Empowerment in this form of democracy is attributed to the capacity development of the citizenry; governance is enhanced through important feedback loops connecting local conditions and values to policy. For Pateman, "only if the individual could become self-governing in the workplace, only if industry was organized on a participatory basis, could ... [individuals] gain the familiarity with democratic procedures and develop the necessary 'democratic character' for an effective system of large-scale democracy" (Pateman 1988, 39).

In the field of green political theory, scholars have built (sometimes rather loosely) on the foundations of these theories of participatory democracy, and also on Schumacher's (1973) vision. In part, this was a response to a Malthusian strain in ecological thinking in the 1970s that focused on the need for a coercive state or even eco-dictatorship and population controls

(Ehrlich 1968; Hardin 1968; Ophuls 1973) in order to halt environmental degradation. The core eco-local arguments focus on the potential for the process of descaling for rehumanizing and revitalizing communities (Curtis 2003; Huckle 2012).

Eco-local theorists are (often sympathetically) challenged by a range of scholars who stress the continued relevance of multilevel governance and a more nuanced understanding of the relationship between the environment and scale (Albo 2006; Bulkeley 2005; Dobson 2005; Dryzek 1992). This is because the conventional understanding of scalar processes relies on the relationship of three distinct units – local, national, and global – either ordered as a ladder-like hierarchy or as a nesting-doll organization of embedded parts. What is missed in both of these accounts is the permeability between units through, for example, international networks of municipalities as well as locally rooted international corporate actors (Bulkeley 2005). Another important recent contribution is a reordering of our understanding of scalar politics to include the ways in which governance may, in fact, emerge from the ground up (i.e., from local practices), though this process is not necessarily one of equally influential or resourced localities. Johnston, Gismondi, and Goodman (2006a, 32–33) highlight the importance of not idealizing the local but focusing instead on multiscaled governance interventions. Put more simply, environmental policy approaches that see the local either as a simple "rule-taker" or as a small territorially bounded unit are incomplete.

In recent years, several other important clarifications and critiques of eco-local thinking have emerged. Murray Bookchin (1999, 286), theorist of social ecology, points out that "decentralism, small-scale communities, local autonomy, even mutual aid and communalism are not intrinsically ecological or emancipatory." One of the most salient pitfalls revolves around the notion of the "local trap" (Purcell and Brown 2005) – the naturalistic association between "the local" and "the good"; that is, that *because* something is local it is necessarily more participatory, more democratic, or more environmentally sensitive. This involves conflating a community of place with community more generally, and especially with a particularly well-functioning or idealized community (Albo 2006; Asiskovitch 2011; Catney, MacGregor, et al. 2013; Marvin and Guy 1997). Authors making this critique tend to point to the myriad challenges both within and between local communities on issues of social justice, distribution, capacity, and resource base. They have also illustrated how the flexibility of localism in terms of its specific application makes it a "protean entity, extremely elastic in its application so that anarchists, social democratic, neo-liberals and environmentalists alike can subscribe to it as a focus for new forms of social organization" (Catney, MacGregor, et al. 2013, 1).

These cautions are appropriate for this research on electricity co-operatives not only because electricity systems in Canada are highly interconnected and geographically spread out but also because ignoring issues of community capacity and distribution risk fetishizing the local. In addition, Carter's (1996) research on green worker co-operatives identifies how many of the assumed environmental benefits of localism and co-operative ownership are often cited rather than empirically demonstrated. Moreover, he argues – pointing to the very significant diversity in forms, intents, and democratic aims of co-operatives – that co-operative models need to be empirically examined in different sectors and contexts. That is certainly not to say that local institutions that support democratic governance are unimportant. Far from it – but provincial, national, and international policy contexts form a fundamental constraint on their development (Albo 2006; Johnson 2009; Lionais and Johnstone 2010; MacArthur 2014). Consequently, what is required is a framework situating co-operative development within broader socioeconomic governance, which looks at the local institution as one part of a wider complex of participatory governance.

Empowered Participatory Governance

Empowered participatory governance (EPG) represents a mode of governance through which electricity co-operatives may contribute to sustainable transitions in the Canadian political economy. It is a framework developed by Fung and Wright (2003) outlining how participatory and democratic institutions can, via development of what they call "countervailing power," further progressive goals within broader structures of powerlessness and lack of democracy. The EPG framework addresses issues such as institutional resilience, networking, and participatory democracy, identified early as crucial elements for addressing the triple crisis. Fung and Wright's work helps address key critiques brought up by critical political economists about the potential for co-optation and neoliberal support in co-operatives, as well as critiques of naive eco-localism. The EPG framework focuses attention on the relative balance of power between different actors within and between institutions and levels within a given system of governance.

This framework also addresses the infamous and perennial reform versus revolution debate in radical political economy. One of the most interesting critiques of co-operatives as an alternative comes from scholars sympathetic to the goal of community ownership or empowerment but cognizant that the co-operative sector has operated throughout Canadian history more as a complement than a challenge to the mainstream economy (Albo 2006; Wright 2010). Indeed, the cited strength in the flexibility of the co-operative movement can also be seen as a weakness – more palatable to the mainstream and thus easily co-opted, more a complement than a challenge (discussed in Chapter 3). Theorists have criticized these movements and the organizations within them (including co-operatives) for supporting the structures of power and exploitation of capitalist accumulation by ameliorating the worst excesses of market society (Fontan and Shragge 2000). Social democracy was seen here as gradualist, focused less exclusively (or not at all) on class, and strategically planned to either overthrow capitalism from within or make capitalism "with a human face," as Žižek (2009) puts it. A co-operative movement silent on the erosion of community power through privatization and continentalism certainly supports this skeptical position.

Fung and Wright (2003) concede that co-optation and participatory window dressing are a possible, but not a necessary, outcome of participatory collaboration. Likewise, Johnson (2009, 2011) finds that it is not simply the structure of participatory institutions that matter but the political context and motivations of elite actors. A variety of mechanisms reduce the advantages of powerful actors through empowered participatory governance. Fung and Wright argue that the more participatory the institutions of governance and the higher the degree of countervailing power, the more empowered (and effective) these alternatives can be in transforming dominant relations. The development of this countervailing power depends on numerous factors, such as the size of the organizational network, its degree of mobilization, and its resource base (Fung and Wright 2003, 260–64). Table 2.1 outlines the breakdown of governance relationship to countervailing power.

What this means specifically for the analysis of electricity co-operatives is that strong networks, political mobilization (both material and ideational), and strong participatory institutions are important. These factors signify the difference between the organizations' being participatory window dressing, and significantly transformational. Indeed, one potential outcome of new initiatives is that they are instrumentally employed as part of a process of communitywashing – employing the term to community market projects with little actual control or local benefit. Ultimately, the political economy of the electricity sector forms a crucial set of constraining

TABLE 2.1
Countervailing power

	Degree of countervailing power	
Governance institutions	Low	High
Top-down administration	I Captured sub-government	II Adversarial pluralism
Participatory collaboration	III Co-optation, participatory window dressing	IV Empowered participatory governance

Source: Adapted from Fung and Wright 2003, 262.

circumstances on whether and how these factors develop (further explored in Chapters 4 and 5). Shragge (1997, 2003) suggests four specific areas co-operatives may make contributions to: democracy, education, alliance building, and mobilization. These factors form the basis of the shift from a pragmatic/reformist tradition in co-operative and community development to a utopian/social change position. Taken together with Fung and Wright's conceptualization of countervailing power, these theoretical contributions form a basic framework for evaluating the transformative potential of co-operatives in the electricity sector.

Local co-operative ownership by itself is clearly not enough to mark a transformative breakthrough. The latter would require local ownership and control over production and distribution, combined with a level of social-movement development. Fung and Wright (2003) develop this idea of supplementing ownership with strong networks and movement-based mobilization, and of moving in this way toward countervailing power and empowered participatory governance. This would involve sustainable economic development, job creation, engaged marginalized populations, and scaled-up actions to push supportive state policy. A further transformative role would involve a shift in environmental consciousness and in the scale and mechanisms of development. Together, these would support a degree of lock-in or "ratcheting up" of effects.

Table 2.2 highlights five elements identified by Shragge, Fung and Wright, and Harvey that are explored in this book to assess the electricity co-operative difference and potential. These elements are also drawn from the grounded

TABLE 2.2

Framework for assessing co-operative potential

	Neoliberal co-operatives (mainstream)	EPG co-operatives (transformational)
Public education[1]	Unimportant	Key priority
Policy impact	Negligible; used to support an increase in marketization and commodification	Significant; challenges existing centres of power in society and the economy
Ownership and control of electricity assets	Partial or minority partnerships	Across distribution and generation, and significant portion of total
Participatory democratic and anticapitalist norms	Unimportant	Central to organizations (solidarity economy)
Networks	Partial or fragmented	Well developed at local, national, and international levels

1 Education, as mentioned earlier in this chapter, can be used in many ways to either challenge conventional practices or to support them (see, for example, Herman and Chomsky 2002). As a result, the *content* and not just the presence of an educative focus deserves attention. The role of education is developed more in Chapters 3, 7, and 8.

research methodology used and described below. This analytical framework focuses on whether electricity co-operatives are (1) consciousness-changing (via, for example, public education); (2) own and control electricity assets (production, distribution); (3) have policy influence; (4) embody a normative set of principles (democratic and participatory) that challenges neoliberal orthodoxy; (5) are networked within and beyond the co-operative movement with other social/economic and environmental justice advocates, and are resilient and creative in times of downturn and crisis.

Summary

If history has taught us anything, it is to be skeptical of political projects making universal and ahistorical claims. There is no inevitable land of freedom and democracy at the end of any one system of governance. Rather, there are only a series of historically specific struggles and contexts, with uncertain outcomes. The actual political and economic constraints on individual agency and collective action, as well as the ideational ones, vary significantly from context to context and matter a great deal to the ultimate role co-operative organizations play (Freire 2000; Žižek 2009). Locally embedded electricity co-operatives may represent an important alternative and are growing in number across the country. These developments are significant given the complexity and scale of the interrelated sustainability crises facing citizens today. So although there may be a key role for them to play in the electricity sector, we must first understand the development and nature of co-operative organizations more broadly: their history, sociopolitical context, and structure.

3

Co-operatives in Canadian Political Economy

Canadian electricity co-operatives form a small but growing part of a broader co-operative sector, with deep historical roots in this country. Examining the ideological and material roots of contemporary co-operation tells us much about the ultimate contribution that this organizational form can make. As Canadian Alexander Fraser Laidlaw argues:

> No co-operative exists in a vacuum but must operate in a given economic and social environment. It must strive, of course, to modify and improve that environment, but it cannot do so unless it recognizes the overriding problems, first of the immediate community, then of the larger region, and finally of the nation and indeed of humanity itself. In the long view the question will be asked: what have these co-operatives and the co-operative movement as a whole done to help people wrestle with the difficulties of life? What is the relevance of co-operatives to the nation's basic problems? (Laidlaw 1980, 51)

Some co-operatives have supported and continue to support dominant governance and policy arrangements as part of a small niche corporate sector. Co-operatives may also challenge these arrangements as part of a broader *movement* with countervailing power to strengthen empowered participatory governance in Canada. The transformative potential of electricity co-operatives thus depends on their intersection within and awareness of specific political economy arrangements.

Co-operatives are diverse and flexible organizations. According to the International Co-operative Alliance, "a co-operative is an autonomous association of persons united voluntarily to meet their common economic, social, and cultural needs and aspirations through a jointly-owned and democratically-controlled enterprise" (ICA 2010). This captures the uniqueness and variability at the heart of the co-operative form: the member-owner link. Co-operatives are not owned or directed by financial speculators, hedge funds, or (inter)national investors; they are owned by their members.[1] Of course, not all co-operatives fit an idealized model of the structures and values. In practice, they vary in their practice of democratic engagement, as well as in their treatment of profit, capital, and commitment to the co-op principles. To understand the role of Canadian co-operatives, we need to understand how – both materially and conceptually – they arose and the specific forms these organizations take.

Co-operatives emerged in times of socioeconomic crisis in Europe – for example, Robert Owen's villages of co-operation in Scotland and the Rochedale pioneers in the 1830s and 1840s. They were founded on the principles that co-operation and reciprocity, rather than competitive markets, formed the natural institutional bases of economies. This echoes Polanyi's later insights that self-regulating markets were a "dangerous fiction" eroding society. Historically, as today, new co-operatives represented a collective response of underserved populations to fill a collective need. They are rooted in redistribution, reciprocity, and use-based economics. They are also rooted in the acknowledgment that markets are imperfect and that collective associations, like co-operatives, were one way to engage in the marketplace in a more advantageous way, either to reshape it or, at a minimum, to bypass concentrated market power.

This chapter traces the contours of contemporary co-operativism in Canada, with a particular focus on the theory and practice of co-operatives' role within neoliberal governance and state restructuring. I outline the distribution of Canadian co-operatives, their potential to contribute to empowered participatory governance, and the challenges faced given the current Canadian political economy context. Despite great potential in this movement, there is a tendency to oversimplify either the form or contribution of co-operatives. Their role in a more progressive political economy hinges on what one is shifting from. In the case of the power sector, moving from public utilities toward more co-operatives and private power actors may not be a progressive change, whereas a restructing from solely investor-owned corporations to more community ownership may indeed result in a more participatory and empowering system for citizens. There are many diverse contributions they may make to communities, but these are dependent on political cultures, public policies, and the membership's political and material (or sectoral) orientation.

Canadian Co-operatives

Co-operative development has a long history in Canada, beginning with the dairy and grain co-operatives of the late nineteenth century, through to the Desjardins development of caisses populaires in Quebec and the Antigonish Movement of the early twentieth century. This last movement blended co-operative businesses with microfinance, adult education, and rural community development in Nova Scotia and formed the basis for the expansion of the credit union movement across English Canada. Co-operatives as a corporate form are virtually ignored in economic and business texts, yet they are startlingly common: four in ten Canadians are a member of at least one co-operative (Government of Canada 2012).[2] In Quebec and Saskatchewan, these numbers are 70 percent and 56 percent respectively (Co-operatives Secretariat 2010b). Some of these organizations are very large; in 2010, Federated Co-operatives had more than $4 billion in assets, 3,100 employees, and an annual revenue of more than $7 billion (Co-operatives Secretariat 2013).

Canadian co-operatives are part of a historic movement stretching back to nineteenth-century cheese factories, creameries, and insurance societies. Canadian co-operatives were leaders in North American co-operation. For example, the Farmers' Bank of Rustico, on Prince Edward Island, was the first people's bank in North America. The first consumer co-op store in North America was established in 1861 in Stellarton, Nova Scotia, founded by British coal miners only seventeen years after Rochedale (Gossen 1975, 44). These early co-operatives emerged to serve the needs of populations that were either underserved or exploited (MacPherson 2009). According to co-operative historian Jack Trevena (1976, 5), Canadian co-operatives formed out of

> the need for people to decrease the power which others held and used against them ... retail co-ops were born because farmers felt a need to act against the power of merchants who charged excessively high prices. Later, credit unions were formed, in part at least, to provide loans at more moderate rates of interest ... Many co-operators deplored the fact that many

business places in their communities, large and small, are owned by people living in far-off places and whom they do not know ... who are alarmed at the extent to which Canada's industries and resources are owned or controlled from outside this country.

The concerns expressed by early co-operators about market power of large companies, local control, security, and nationalism echo those of many Canadians today (Bradford 2005; Neamtan 2002; Teeple 2000).

Fascism, state socialism, and the advent of the welfare state led to more centralized economic systems and a correspondingly diminished role for co-operatives in some sectors. In fact, throughout the long history of co-operatives, they have developed in cycles, with significant development during times of crisis, then ebbing when key needs are met. There was, for example, a surge in co-operative development in Canada between the 1930s and 1940s. In 1933, there were 686 co-ops with 325,369 members. By 1944, these numbers had more than doubled to 1,792 with 690,967 members (Faucher 1947, 188). As earlier larger co-operatives merged, consolidated, and settled into the broader economy, many co-operatives became more of a sector rather than an oppositional and transformative movement. Today, there is a resurgence in co-operative development across Canada and around the world as communities seek to respond to austerity measures and to challenges of sustainable development not being tackled at the national level. Although this may reflect a natural progression of social movements along a life cycle (Gamble et al. 2007), it supports the view that co-operatives are a source of systemic and transformative response to neoliberal globalization.

Types of Co-operatives
Contemporary Canadian co-operatives are diverse organizations. They can be asset rich or poor, small or large, and technologically complex or simple, and they can exist across an extremely wide range of sectors (Fairbairn 2003; Fulton 1990; MacPherson 2009). These sectors include dairy marketing, petroleum refining, lumber, financial services, coffee roasting, health care, car, and insurance services. Despite the significant diversity that exists between these organizations, five main types exist based primarily on the nature of the member link to the organization: consumer, producer, worker, financial, and multistakeholder/solidarity (see Table 3.1). The member link is one of the key factors differentiating co-operatives organizationally from investor-owned actors in the private sector, as it connects those that control the organization with an interest in the good or service itself, rather than in profit alone (if even at all).

TABLE 3.1
Types of co-operatives

Members	Description	Canadian examples
Consumer	Members are the customers (of both goods and services) of the business	Mountain Equipment Co-op, Federated Co-operatives
Producer	Members use the co-operative to sell or market goods; also sell business inputs to members	Agropur, Gay Lea
Worker	Members are the employees of the business	Girardville Forestry Co-operative, Sustainability Solutions Worker Co-op
Financial	Members are the customers of the financial institution	Desjardins, VanCity, The Co-operators
Multistakeholder/ solidarity	Newer form; different members all form the co-operative: workers, service users, and locals, e.g., health care and tourism co-ops	La Corvée, Co-op de solidarité en soins et services de St-Camille

Source: Adapted from MacPherson 2010; CCA 2010.

Whereas worker co-operatives are formed to meet the needs of employees – including good-quality, stable work – producer co-operatives are aimed at helping farmers and craftspeople market, sell, and distribute their goods. Consumer co-operatives operate in order to supply goods and services that populations are otherwise unable to secure. Retail co-op stores across western Canada are often the only local grocer or hardware store in rural communities (MacPherson 2009). Solidarity co-operatives – multistakeholder co-ops that bring together producers, consumers, workers, and members of the broader community – have arisen predominantly in Quebec. According to Quebec's Direction du développement des coopératives, there were 549 registered solidarity co-ops in the province in 2011, an increase of 124 since 2004 (Ministre du Développement économique, de l'Innovation et de l'Exportation 2011). Their members come from many different member groups (consumer, worker, stakeholder) and are unified by a similar goal; for example, health care workers might form a home-care solidarity co-operative with elderly patients and broader community members.

Co-operative Assets and Membership

The material strength of Canadian co-operatives – in terms of assets, membership, and geographic purchase – matters for the ultimate potential of the sector to contribute to empowered participatory governance (see Chapter 2). Co-operatives form a substantial and growing part of the Canadian economy. The latest available report on co-operatives in Canada illustrates the contemporary size and scope of the co-op sector in Canada: there were 7,839 financial and non-financial co-operatives in 2010, growing from just 1,100 in 1930. Co-operative membership has also grown over this time period, from 756,000 to more than 17 million, outpacing population growth in the country by a multiple of three (Co-operatives Secretariat 2010a, iv; Government of Canada 2012).[3] Co-operatives in 2010 employed 150,000 people and had assets of $330 billion; however, 48 percent of nonfinancial co-operatives had no paid employees and operated solely with volunteer labour (Government of Canada 2012, 5; Industry Canada 2015, 8). Despite nonfinancial co-operatives being five times more numerous, they hold only one-tenth of the assets of financial co-operatives (Co-operatives Secretariat 2010a). Appendix 3 provides a summary of the activities of the top ten nonfinancial co-operatives in Canada as of 2010.

Provincial Distribution of Co-operatives

Co-operative membership and penetration differ across the country, with the majority (by number of co-operatives) located in Quebec and Ontario. Table 3.2 illustrates the number of co-operatives by province in 2010. Co-operative membership (as a percentage of population) is strongest in the

TABLE 3.2
Co-operatives by province and membership, 2010

Province or territory	# co-ops	% of total co-ops	Members[1]
BC	301	6	3,697,600[2]
AB	411	8	1,184,400
SK	579	11	483,400
MB	224	4	426,200
ON	708	14	143,100
QC	2,379	47	1,249,800
NB	101	2	88,700
NS	284	6	44,200
PEI	54	1	17,000

▶

◄ **TABLE 3.2**

Province or territory	# co-ops	% of total co-ops	Members[1]
NL	19	0	39,200
YT	6	0	9 (2007)
NWT	25	0	7,891 (2007)
NU	31	0	20,506 (2007)
Total	5,094		7,397,600

1   Member data is based on reporting of co-operatives on the annual co-operative survey conducted by Industry Canada (and formerly by the Co-operatives Secretariat). In the 2010 survey, these made up 64 percent of co-operatives (5,094 of 7,865) in Canada in 2010 (Industry Canada 2015). As membership from reporting co-ops from the territories is aggregated (24,000 in 2010) in the Industry Canada data, figures in the table were drawn from the previous 2007 survey (Co-operatives Secretariat 2010a).
2   Mountain Equipment Co-op is headquartered in British Columbia; all national members are included in the BC figure.
Sources: Co-operatives Secretariat 2010b, iii; Industry Canada 2015.

Prairie and Atlantic provinces (Co-operatives Secretariat 2011). Membership numbers for British Columbia are inflated because Mountain Equipment Co-op membership (with more than 2 million members in 2007; up to 3.3 million in 2009) is accounted for in the data for British Columbia, whereas in practice its membership is spread across the country. Nonfinancial co-operative numbers show the prevalence of co-operatives in smaller and more rurally based provinces, particularly where co-op stores and farm co-ops play a large role (Industry Canada 2015).

The Co-operative Difference

The co-operative difference is used not only to justify academic scholarship on co-operative development but also to encourage state agencies and policy makers to recognize the value of these organizations to key policy goals such as employment, training, social cohesion, and rural development, and to support them financially and legislatively (Vaillancourt 2008). Co-ops can be differentiated from other businesses on numerous fronts (see Table 3.3). These differences are both material (legal form/structure) and ideational (co-operatives' values, ideals), and are embodied in the ways in which profit, control, and membership are constituted in the organization.

Three elements of the co-operative difference are addressed below: international principles, profit, and democratic constitution.

TABLE 3.3
Co-operative and business comparison

	A co-operative	Private, investor-owned business
Profit	Surplus refunded to members in proportion to patronage.	Surplus allocated in proportion to investment.
	Surplus earnings or profits belong to members, distributed at annual meeting, yearly as recommended by board.	Surplus earnings or profits belong to the corporation, distributed by board of directors.
Control	A co-operative is a system that guarantees Canadian control of Canadian enterprise.	Constant vigilance is needed to prevent takeover of Canadian business and industry by foreign interests.
	Ownership is in the hands of its members in the community who use the service.	Ownership is in the hands of investors.
	Its control is democratic; each member has one vote.	Control is unequal, by majority of shares.
	Shares are held in the name of members only and are not traded for speculation.	Shares may be freely traded and fluctuate in value.
	Proxy control is rare.	Proxy control is commonplace.
Organization	An organization of users. Essentially a union of persons.	An organization of investors. Essentially a union of capital.

Source: Gossen 1975.

Principles

The differences between co-operatives and shareholder-owned businesses arise in part from the former's federation in the International Co-operative Alliance and the international co-operative principles (revised in 1937, 1966, and 1995) developed by it. These nonbinding principles are based on those of the 1844 Rochedale movement in England and work as rules of conduct "subject to interpretation" (Bergen 1984, 184). The principles, presented fully in Appendix 4, are (ICA 2010):

1 Voluntary and open membership
2 Democratic member control
3 Member economic participation
4 Autonomy and independence

5 Education, training, and information
6 Co-operation among co-operatives
7 Concern for community.

The international principles of the co-operative movement serve as guidelines to assess whether any given organization is working within the remit of the movement more generally. They form a key part of the multiple bottom-line approach of the movement, which prescribes a lens of "people before profit," and the integration of social (member and community) concern as a fundamental part of the organizational form. From time to time, the principles are reassessed and revised to take account of new challenges, and to allow for more flexibility. The most recent revision of principles in 1995 was chaired by Ian MacPherson and focused on the challenges that globalization presented to the co-op movement. The 1995 revisions added a seventh principle that focused on the broader community and sustainability. Other changes included adding gender as one of the types of discrimination to avoid, removing a limitation on interest to share capital, and reintroducing the principle of independence present in earlier manifestations but absent in the 1966 principles.

The International Co-operative Alliance serves as a space for discussion and debate between international co-operators over principles and provides a focal point for interacting with international bodies such as the United Nations. It has been playing an important coordinating role at the international level, liaising with national state and co-operative actors for more than half a century. In 1946, the International Co-operative Alliance, together with the World Federation of Trade Unions, the International League for Human Rights, and the International Chamber of Commerce, were the first NGOs assigned consultative status at the United Nations. National and international mobilization led to 2012 being declared the International Year of Cooperatives by the UN General Assembly, raising the profile of these organizations worldwide. In 2009, the Canadian government ratified this move, and in 2011, the US Senate followed suit (though, as we see later in this chapter, in 2013 the Conservative government significantly reduced targeted support for the co-operative sector). The networking and movement building that takes place within the International Co-operative Alliance is important, as it provides a space for strengthening the degree of countervailing power that these organizations can exert on policy makers.

Profit

Co-operatives are voluntary organizations that often operate in the market as social enterprises. Profit is not the purpose of a co-operative. Service to members – as defined by those members – is, so co-operatives can also be nonprofit organizations. Co-operatives differ significantly from other businesses in that their treatment of profits can be structured as either for-profit or not-for-profit. They key difference between these two (other than tax implications) is that for-profit co-operatives distribute their profit – "savings," in co-operative language – back to their membership. According to Art Postle, former CEO of Federated Co-operatives Limited (FCL), there are benefits "from coming together as a co-operative because it gives you some benefit of availability of supply, cost of supply, quality of product, all of those sorts of things. But you don't form one from the perspective of 'we can make an investment in a co-operative'" (personal interview, July 8, 2009).[4]

The treatment of profits and member orientation leads to a flow of capital and skills back to the local communities, which in turn leads to local economic development benefits over other corporate forms (Fairbairn et al. 1995; Ketilson et al. 1998). The role of co-operatives in rural (and particularly northern) development has been well documented in Canada (MacPherson 2009). Where co-operatives engage in lucrative business areas, as with Consumers' Co-operative Refineries Limited (CCRL) and oil refining, the returns are circulated back through the membership and into Canadian communities. For example, since the CCRL is wholly owned by Federated Co-operative retail stores, these rural firms continue to operate supported by funds from the refining side of the organization. In fact, according to FCL, an umbrella co-operative for retail co-op stores in western Canada, it returned $355.7 million in patronage refunds to members on 2010 FCL purchases.[5] This included $207.3 million in cash equity to individual members, and more than $1 billion cash in the past five years (FCL 2011).

The treatment of profit within co-operatives also leads to local development in other ways. First, when primarily locally financed, they don't need to come up with market rates of return for investors (often in excess of 10 percent). According to one co-operative developer in Quebec's Lower St. Lawrence region, in the forestry sector, the only mills open and functioning now are co-operatives, because they can function with 3 percent returns (Gagnon personal interview, May 16, 2010). Second, a 2008 Quebec Ministry of Industry and Commerce study found that co-operatives have almost twice the survival rate compared with other businesses in Quebec, with the co-operative advantage (the gap between the two rates) growing as time goes on (see Table 3.4).

The lack of prioritization of profit for profit's sake in the movement may make these organizations more likely to take other values, including environmental ones, into account (Gertler 2001). Even if sustainability is

TABLE 3.4

Co-operative and business survival rate in Quebec

	After 3 years	After 5 years	After 10 years
Average survival rate of co-operatives	75%	62%	44%
Average survival rate of Quebec companies	48%	35%	20%
Co-operative survival advantage	1.56 times greater	1.77 times greater	2.2 times greater

Source: Adapted from Clément and Bouchard 2008, 22.

explicitly listed in the co-operative principles, the role and circulation of profit, together with a local connection to resources (Ostrom 1990), may play a key role in better positioning co-operatives to help address future challenges. Co-operative association employees in Ontario and Nova Scotia respectively put the issue this way:

> If the co-op when it starts understands people before profit, [sustainability] is an easy sell if they aren't there already [involved in green activities]. The green solution doesn't get reduced to "Oh God, it's so expensive to buy recycled paper" ... the resurgence of the local food movement has been hugely co-op centric. Most are saying, "Why would you want to do this any way other than farmer-owned or consumer-controlled or both?" (Former co-operative developer personal interview, July 20, 2009)

> I think there is always the self-interest that the co-operative sector captures – that is, I don't want to pollute my own community, you've got peer pressure, you've got other things. It maybe shifts the decision making on a social level. (McLelland personal interview, May 18, 2010)

Democratic Governance and Accountability

Members control co-operatives through equal voting rights, and each member, regardless of financial investment or use, gets an equal vote. Decision-making power is not based on how many shares or how much capital you have. This particular allocation of voting power is one of the reasons advocates for co-operatives argue that co-operatives are more democratic than other businesses. Member control of co-operatives also means that, in general, the businesses operating in the community are subject to the population that owns and patronizes them. The people in the co-operative sit on the school boards, are involved in local politics, and live in the same geographic area. As a result, the interests represented in the co-operative are not solely (or even partially) that of profit maximization but often revolve around service provision, employment, and identified member needs.

As explored in Chapter 2, the co-operative governance difference is further supported by the significant informational and coordination advantages to structuring institutions locally and collectively (de Peuter and Dyer-Withford 2010; Ostrom 1990; Rees 2010). For instance, organizations where resource users themselves have a say and stake in their governance are better placed to adapt to local conditions and complex policy challenges (Bradford 2005). In co-operative parlance, the emphasis is on stakeholders, not shareholders. The theorized benefits of closing the user-owner loop and tightening feedback between managers, owners, producers, and consumers is that adaptations can be made more quickly and that the users have incentives to ensure sustainability of their livelihoods while the requisite social networks deepen compliance.

Control of local retail co-operatives, housing, and workplaces can affect the availability of supply, cost of supply, and quality of product. This control is, of course, bounded significantly by the state regulations and the political economy within which co-operatives operate. Indeed, the value of co-operatives is often in responding to broader economic conditions and problems. Several examples illustrate the impacts co-operatives can have in this regard. In Quebec, co-operative ownership prevented the closure of lumber mills, as the membership prioritized secure local employment over the profit margins private companies require to stay in operation (Gagnon personal interview, May 16, 2010). In Saskatchewan, the Consumer Co-operative Refineries Limited produced a unique diesel product (EP3000) at the suggestion of its farmer members, with more energy and miles per gallon (Postle personal interview, July 8, 2009). In Ontario, the Everpure co-operative supplied biodiesel to local residents, keeping prices competitive with gasoline by buying in bulk. Similar examples of service or product innovation are found in daycare, housing, and a wide range of other sectors (Deller, Hoyt, and Hueth 2009; Quarter 1992; Restakis 2010).

The direct and local connection co-operatives embody between the management and the users thus provides more stability, and resilience from the vagaries of businesses trying to maximize a bottom line for shareholders. The stability arises from the fact that co-operatives are less likely to cease operations when profit margins shrink, as the member interests are only partially financially driven. According to Hammond Ketilson, it was through the use of democratic structures and processes that co-operatives survived where other business could not, developing and using social cohesion to mobilize scarce resources (Fulton and Ketilson 1992; Ketilson et al. 1998).

Co-operative development also leads to information and capacity building at the local level. According to one Nova Scotia Co-operative Council employee, "it is always good to have competition if you can have local competitors ... that brings a good dynamic to a community, whether it is [in] food processing or energy generation; if someone is inflating their price, they can be undercut; that knowledge is there" (McLelland personal interview, May 18, 2010). Hence, the development of local alternatives, of having locally controlled agents, skills, and options, makes a difference in the capacity of memberships and the communities they live in to address market failures.

Challenging the Co-operative Difference

Despite the many contributions that co-operatives can and do make for Canadians, thorny issues persist over how deep the co-operative difference goes toward creating empowered forms of governance. These issues include the role of investment, lack of movement building, member participation, and adherence to social and environmental bottom lines. At the heart of these challenges is not the potential of the co-operative structure itself but the actual practice of boards and members. Co-operatives (particularly for-profit ones) are similar to private-sector corporations insofar as they operate in the marketplace, selling goods and services to the public. They are owned and governed by private individuals, not the public at large, and are in many ways firmly part of the private sector. This is particularly the case with larger co-operatives that work as bulk purchase arms, as the member-governance link is diffused through the organization. In response to these challenges, a new co-operativism is emerging that seeks to move beyond some of these challenges to reinvigorate the solidarist and anticapitalist underpinnings of co-operativism (Vieta 2010).

I am not alone in raising these complexities (Asiskovitch 2011; Graefe 2006; Stanfield and Carroll 2009). Kasmir's (1996) study of the Mondragon co-operatives and the working class in the Basque region of Spain, for example, warns against idealizing co-operatives and ignoring the multiple and multilayered forms of exploitation that can take place within them.

Avi Lewis, director of *The Take*, commented in a book on co-operatives and capitalism that "the global co-operative movement could provide a genuine alternative to the ravages of predatory finance capitalism – if only it started acting like a movement!" (Restakis 2010, i). Put simply, Canadian co-operatives have developed, from their early days as part of a social movement to their current standing with a more established role in the economy. As this development has taken place many have professionalized, narrowed their social roles, minimizing the kind of challenge they can pose. It may, in fact, be time to build a new co-operativism (Vieta 2010).

Member engagement is a limiting factor for co-operative democracy and accountability. It is possible to be a member of a co-operative in name only, with little understanding of your rights or of the co-op structure itself. Low member engagement leads to challenges of staffing volunteer boards and bringing new generations into co-operative development and participation. A 2010 Ipsos Reid poll on co-op awareness in Canada found the following (Ipsos Reid 2010):

• 59 percent of respondents were "not at all" or "not very" familiar with co-operatives. Only 7 percent reported being "very familiar."
• One in five respondents was actually a member of a co-operative.
• 44 percent did not know the difference between co-operatives and other types of businesses.
• Respondents were confused as to which businesses were co-operatives. When provided with a list of organizations, 49 percent thought Costco was a co-operative, 40 percent thought WestJet was, and 35 percent thought CUPE was.

According to the manager of an Albertan energy co-operative, "we're not promoting the co-operative spirit. I don't care, we just aren't. Unfortunately, it'll probably take a catastrophe, and usually it's the small guys that suffer when that happens" (personal interview, December 1, 2009).

Challenges of representation also persist in these organizations. The governance issues within co-operatives, particularly the relationship between boards and the membership, is often highlighted in the literature, as is the role of volunteers in co-operatives and the treatment of nonmembers (workers and other community members) (Fulton 2001; Laycock 1990; MacGillivray and Ish 1992; Restakis and Lindquist 2001; Turnbull 2007). For example, professional managers and boards can manage and steer information flows in ways that the membership is unaware of, or unengaged with. This is not a problem unique to co-operatives, but it certainly affects the depth and veracity of any claims in a specific co-operative to engendering economic democracy at a broader level for communities. Addressing the localization and community representation argument seriously means dealing with the sticky issues of political ideology that have plagued the co-op movement from day one: Is it conservative or solidarist? In the United States, according to Paul Gipe, a community power expert based in California and an OSEA policy adviser, "rural electricity co-operatives don't necessarily practice the principles; they don't necessarily believe in democratic participation. [Members] do have a vote for the board because that's in their bylaws, but [the co-operatives] don't necessarily believe in that" (Gipe phone interview, April 7, 2010).

Neoliberal governance is setting the context for a revitalization of co-operativism in Canada as provinces retreat from services provided and developed in the welfare-state era. New co-operatives are being pushed from the grassroots level of material need, as well as from policy from the state and transnational organizations. This state support raises both challenges and opportunities for co-operatives, as the specific policy intention varies greatly between states – from the support for an expansive co-operative economy in Venezuela to a marginal neoliberal role in Britain and Canada. If anything should be clear from this chapter so far, it is that co-operatives are no one thing; they are fluid and flexible, and reflect the societies within which they operate. Co-operatives, as creatures of their member needs, are affected in role, nature, and number by state policy and larger shifts in political economy. Public-policy choices can either strengthen or weaken the co-operative difference. Neoliberal policies that devolve responsibility for social-service provision to co-operatives, without the requisite funding or policy coordination, ensure that initiatives remain gap-filling and partial.

Co-operative leaders are not necessarily unaware or sanguine about these developments, however (Mayo 2011). Crucial tensions thus emerge, as they always have, between short-term provision of needs and long-term socioeconomic transformation and change. Without an institutional program to enhance the progressive and democratic aspects of co-operativism, existing centres of state power will shape them in a neoliberal direction (Golob, Podnar, and Lah 2009; Graefe 2006; Keevers, Skykes, and Treleaven 2008). Clearly, the latter is taking place both in Canada and in the United Kingdom. Co-operatives play a role in shaping that context, but the emphasis on independence and self-help in the co-operative literature does little to advance countervailing power. At the same time, elements within and around the co-operative movement –in labour, social, and solidarity organizations – and in progressive political circles are forming a response. The new co-operativism is a rearticulation of the movement, enhancing solidarity and progressive political struggles. Vieta (2010, 2) notes: "Even with the entrenchment of neo-liberalism over the past four decades, co-operative practices and values that both challenge the status quo *and* create alternatives to it have returned with dynamism in recent years."

Co-operatives and Public Policy

One of the reasons co-operatives are resurgent today across the provinces is that they contribute positively to local economic development (Deller, Hoyt, and Hueth 2009; FCPC 2013; Government of Canada 2012). Another is that they straddle two problematic dichotomies for policy makers: public (state) and private (market), left and right. Co-operatives are remarkably flexible institutions and suit the political leanings of both the ideological right and left. Being in the private sector, they allow states to remain rhetorically committed to letting the private sector operate with limited intervention from the state. The co-op movement itself is ideologically rather agnostic. This provides a useful opening for state agencies seeking to marketize public services (Restakis 2010; Restakis and Lindquist 2001). The core values can certainly fit within a neoliberal context. This flexibility is a short-term strategic asset, but in the long term it is problematic for the kinds of contribution the social economy and co-operatives can ultimately make (Graefe 2006). Canadian co-operators made similar points. Says a former co-operative developer within Ontario:

> We've really benefited from the ability to look past ideology. There's a fair bit of us that have very strong ideological viewpoints ... The CCF was left wing ... but lots of the Social Credit guys formed Federated Co-ops. The key is that the ideology doesn't necessarily come along with the model and the ability to see the commonalities has done the co-op model well. (Personal interview, July 20, 2009)

One Albertan biofuel co-operative developer described the politics of his co-op members as follows:

> Most of us are right wing. We're not an old line co-op either. I think my grandfather in 1905 was the first president of the Killam co-op store ... he came out of Iowa to here. I'm sure he was the first chairman, formed a co-op

and brought barbed wire in ... I wouldn't vote NDP if my life depended on it. But we shop at the co-op all the time. (Personal interview, December 1, 2009)

As a result of the political diversity within the co-operative movement, the New Democratic, Conservative, and Liberal parties in Canada support them, albeit to differing levels, as discussed below. There was a range of responses from interviewees regarding the role political parties played in aiding co-operative development, with roughly half arguing that it doesn't matter which party is in power when it comes to facilitating co-ops. One provincial co-op association employee noted:

The [Nova Scotia] Conservative government was also very supportive. There hasn't been a big change, really. It is maybe like farmers' markets; no matter how political you are, it is hard to argue against them. Here we've had a good relationship with both. I wouldn't say co-ops have anything to fear with an NDP government; when the government was in opposition, if there was a grant to Michelin, that would be criticized. There's not the same dynamic about co-operatives. (McLelland personal interview, May 18, 2010)

But for others, the NDP was clearly more supportive:

In 1992, when the NDP was in power in Ontario, we negotiated an agreement with them to support worker co-operative development, and we had five offices across Ontario with ... regional developers [and] a development strategy process with prefeasibility through formation ... It was fantastic. When the NDP didn't get elected, all that was swept away, we lost all our momentum. (Personal interview, July 23, 2009)

The relationship of co-operatives within broader processes of accumulation and legitimation is complex. Co-operative activity is, on the one hand, indicative of co-operative success at filling and meeting the needs of a population. From another perspective, the fact that those needs exist in the first place, or are increasing, means that other elements of the political economy picture are eroding the security of populations, so that co-operative development may actually be a symptom of deeper problems in a given political economy.

Public policies and state agencies at both the provincial and federal levels support co-operative development.[6] State policy and programs, together with the work of co-op associations and the communities they locate in, explain the diversity of co-ops across the provinces. Appendix 2 provides an overview of the relevant co-operative agencies, legislation, and tertiary co-op associations at the federal and provincial levels.

Federal

At the federal level, significant changes have taken place in recent years that may have negative effects for the co-operative sector. For example, the 2011 federal budget amended RRSP regulations, prohibiting RRSP eligibility for share ownership of 10 percent or more. This issue directly poses a challenge for capitalization of co-operatives, particularly worker co-operatives (FCPC 2013; Government of Canada 2012). In 2012 – ironically, the International Year of Cooperatives – several further changes were announced in the budget. According to the 2012 federal government *Status of Co-operatives in Canada* special report, these changes included cancelling the long-standing Co-operative Development Initiative, reducing the staff of the Co-operatives Secretariat in Agriculture and Agri-Food Canada (AAFC) by more than 85 percent, and failing to establish new funding support (Government of Canada 2012, 43).

Responsibility for co-operatives at the federal level was transferred to Industry Canada. This move was supported by numerous actors within the co-operative sector as being more reflective of the diversity of co-operative services: "Several witnesses, including the Coop fédérée, emphasized that it was now inappropriate for RCS [the Rural Co-operatives Secretariat] to be under AAFC's aegis since co-operatives no longer have a solely rural or agricultural purpose" (Government of Canada 2012, 49). However, in the review, concerns were also expressed by interviewees about a lack of focused attention or targeted staff to support co-operatives going forward.

Subsequently, the Canadian Co-operative Association (CCA) and Conseil canadien de la coopération et de la mutualité voted to create a new apex organization for co-operatives in June 2013 to provide enhanced support for the sector: Co-operatives and Mutuals Canada, which "laid the foundation for a stronger, more sustainable bilingual association for Canada's co-operative and mutual enterprises" (Leblanc and Guy 2014). This organization now represents Canadian co-operatives within the International Co-operative Alliance.

Provincial

When comparing provincial government institutional support for co-operatives, four provinces stand out: Quebec, Nova Scotia, Manitoba, and most recently, Newfoundland (Table 3.5). Unlike the rest of the provinces in Canada, where co-operatives deal with broader business development or

TABLE 3.5

Provinces with specific co-operative government agencies

Province	QC	NS	MB	NL
Department/ Policy	Direction des Co-operatives Investissement Québec (Régime d'investissement Coopératif)	Co-operatives Branch, Access Nova Scotia	Housing and Community Development (Cooperative Development Services)	Newfoundland and Labrador Registry of Co-operatives Department of Business, Tourism, Culture and Rural Development (Co-op Zone project and a Regional Co-op Developers Network)

Source: CCA 2011b.

financial regulatory agencies, in these four provinces, targeted support (staff and funding) has been set up specifically for co-operatives. In three of these cases, the provincial co-operative associations do the up-front development work, funded by a provincial department.

Strong co-operative development systems, like those in Quebec and Nova Scotia, lead to a sector that is better resourced, better networked, and better able to push its agenda into many parts of the economy. By far the most advanced is the province of Quebec, which funds and administers a co-operative development program. This program works in conjunction with the Fédération des coopératives de développement régional (CDR) du Québec to support twelve regional CDRs across the province. A 2007 initiative in Newfoundland used a similar model wherein the Department of Business, Tourism, Culture and Rural Development partnered with the provincial co-op association to fund and administer a program called Co-op Zone. Through it, nine regional development staff facilitate, educate, and promote co-operative development in the province.

In Nova Scotia, the Co-operatives Branch of Access Nova Scotia works with the Nova Scotia Co-operative Council to help fund and promote co-operative development in that province. The province of Manitoba's

Cooperative Development Services branch provides help (e.g., on incorporation and on navigating regulations) and, in 2010, the housing and community development minister announced a tax credit for donations of up to $50,000 to a new co-operative development fund (Government of Manitoba n.d.). Until 2007, Saskatchewan was on this list of supportive provinces, as it had a department for Regional and Co-operative Development, but with the 2007 election in that province, development and business fell under the remit of a new agency, Enterprise Saskatchewan, and co-op–targeted organizational support ended.

Neoliberal Co-operatives?
The contemporary resurgence of policy interest in co-operatives arose in response to the processes of state restructuring. This interest has surged since the global financial crisis of 2008, with states looking to cut budgets and find new and cheaper ways to provide services that the public has come to expect, but with reduced revenue. Until recently, the development of Crown corporations and provision of public electricity, education, social security, and health care all served to reduce the need for co-operatives. According to MacPherson (2008, 636), "to a significant degree, the historic communitarian instincts of the international movement were blunted by the advent of the welfare state." One interviewee in Alberta articulated the tension and connection between the co-op–public-ownership relationship this way: "If you have public, why would you even have co-op? It really is one big co-op, only the government controls it: a co-op of the whole province. It is the same concept [as co-operativism], we're all owners because we all own everything in the province" (Nagel personal interview, November 27, 2009).

The political variability and flexibility of the co-operative form, together with the very real contributions to social and economic development based on the co-operative difference, make them an appealing vehicle and useful ally for neoliberal states. Further, using co-operatives as service delivery mechanisms, as Alberta did with gas co-operatives, adds a level of legitimation to the withdrawal (or even refusal to enter) service provision. One business developer at the Nova Scotia Co-operative Council put it this way: "To be honest, politically, there's some sectors of the economy that, if a co-op becomes involved – [for example,] health care – there's probably less of a backlash against privatization and big bad government coming in" (personal interview, May 18, 2010).

Co-operatives and organizations like them can thus play a flanking role for neoliberal policy, essentially cleaning up after and supporting neoliberal processes (Jessop 2002). With basic needs met by pooling and community action, the most profitable parts of the economy are left to the private sector, and in so doing, "self-help" erodes broader progressive goals. Being consigned to the flanking role also leads to dangers for co-operative survival and the co-operative contribution. For MacPherson,

> the withdrawal of the state from so many activities opened opportunities for new co-ops, especially health and social co-ops. The apparent triumph for liberal, capital-driven firms seemed to some to beg other challenges: how were co-operatives different? Were successful co-ops destined inevitably to become private firms, "demutualized" to pursue market advantages and profit for small groups if not their entire memberships or their communities? (2008, 639)

The experiences of UK co-operators suggest cautionary lessons for their counterparts in Canada (Jordan 2010). The enthusiasm for co-operatives in Britain has come from both Labour and the Conservatives, keen to appropriate the volunteer time, energy, and skills for social cohesion while keeping the services off the state ledgers by contracting out public services to private organizations (Hunt 2010). The UK co-operative association has urged co-operators to be cautious of how the movement is being used to justify public austerity measures. In a 2011 press statement, Ed Mayo, secretary general of Co-operatives UK, argued:

> We are wary of some elements of the government's approach to opening up public services to outside providers ... First, there are serious issues facing public sector employees and users looking into the co-operative option – from uncertainty about jobs and pensions to the challenge of public sector workers setting up new businesses – that need to be addressed if public sector mutuals are to succeed. Second, in the current context, it won't help staff or users if all the government does is to open the door to privatisation with fake mutuals that fail basic quality tests of member ownership and democracy. (Mayo 2011)

Summary

Co-operatives are no institutional panacea, but they certainly can play an important role, along with labour unions, the solidarity economy, and more conventional political agents, in working toward the creation of what Hunt (2010) calls a "polyculture of dissent." The co-operative model, despite its challenges and problems, still represents a unique and potentially more democratic organizational form. The institutionalization of closer feedback loops, transparency, member voting, and economic development support a "co-operative difference" from shareholder-owned firms. With co-operatives' roots in solidarism, collective action, and early social movements, the potential for their contributing to a more progressive political economy exists. The significant purchase that these organizations have in the Canadian economy, together with the lack of attention paid to them, makes them an important area for study. So too does the recent re-emergence of policy initiatives aimed at contracting out services to co-operatives.

A sober analysis of co-operative development and potential, however, needs to go beyond cheerleading and advocacy of these interesting and unique institutions to address critiques that the movement represents a gentler form of capitalism and is eroding the welfare state from the left. Co-operative values can be read as either conservative or progressive, depending on which ones you highlight. Many co-operatives that have existed for years are not part of international solidarity movements and are openly hostile to these broader movements to repoliticize co-operatives. The co-operative difference does not always lead to preferable outcomes for social justice advocates. It is, however, strengthened by solidarist projects, by political mobilization, and by a focus on participation and public education within certain sectoral contexts. The next two chapters examine the various drivers reshaping electricity systems in Canada, where renewable energy co-operatives are developing.

4

International Forces for Power-Sector Restructuring

Much as an understanding of Canadian co-operativism matters for analyzing the impact of electricity co-operatives, so too does an understanding of the changing political economy of electricity generation and distribution. The electricity sector is undergoing momentous change and, as Ted Craver of Edison International argues, faces "more change in the next ten years than we've seen in the last hundred" (quoted in Achenbach 2010, 138). These changes include a restructuring of grids, of ownership structures and governance arrangements, and the introduction of new technologies. Electricity reforms are required to mitigate global climate change and to provide basic human services (Devine-Wright 2011; Jacobsson and Lauber 2006; Scheer 2007). How, why, and where power-sector reforms take place influence the potential for electricity co-operatives to contribute to more sustainable and empowering forms of governance. This is because co-operative feasibility is shaped by the macropolitical context within which the possible is set. Currently in Canada, a neoliberal frame sets the boundaries for what is financially possible and politically imaginable, and for likely future policy trade-offs. Despite the rhetoric of open competition that pervades neoliberal reform models, real cultural and material constraints exist in the electricity sector in the way in which particular actors – in this case, co-operatives – are treated.

The complex forces driving Canadian restructuring arise from both domestic and international sources. The international drivers of power-sector restructuring are addressed below. This chapter outlines the overarching forms and forces of reforms, with a particular focus on the United States and continental pressures on Canada. This is important because neoliberal reforms in Canada, explored in Chapter 5, are part of a broader project of electricity market restructuring that has taken place in many jurisdictions around the world – for example, in the United States, the United Kingdom, India, and Chile – since the 1980s, often with problematic results. These changes cannot be explained simply as a function of domestic financial or technological challenges. Rather, these reforms are being pushed by a range of actors in governments, international organizations, regulatory bodies (for instance, the Federal Energy Regulatory Commission), and corporations seeking to open public services to private accumulation (Blue 2009; Cohen 2004; Slocum 2001). These pressures spill across international borders and create a climate for states that have yet to conform to start adapting marketizing processes and institutions for their power sectors.

Power-sector restructuring has manifested in different ways in various places. In some countries, including Chile and the United Kingdom, restructuring involved outright privatization. In other cases, as in California, the primary mechanism was deregulation. The incumbent fuel sources, domestic political coalitions, and governance structures in different states have led to a patchwork of power sectors internationally wherein both public agencies and private actors play important roles (Victor and Heller 2007). In each of these diverse national experiences, restructuring has come with significant controversy and cost (Anderson 2009; Dubash and Williams 2006). For Canadians, developments in the United States are of particular interest, since the two countries are so closely linked by ever-deepening trade ties. Indeed, according to Howse and Heckman (1996, 134), a "combination of American regulatory activism, Canadian regulatory inertia, international trade law rules, and Canadian interest in continued access to American market may bring about an integrated Canada/US [power] market." Continental integration and restructuring presents challenges for the development of a greener or more democratically constituted power sector; these include increased power trading and generation for profit, and diminished incentives in restructured markets for demand management.

The What, Why, and How of Electricity Restructuring

Power sectors are structured along a continuum of integration and marketization. At one end is a service-oriented system wherein one actor (typically, a public utility or regulated private monopoly) is responsible for all aspects of power delivery within a given jurisdiction, from generation, through transmission and distribution, to retail. At the other end is a functionally unbundled market-oriented sector with competition and actors at each point in the generation, transmission, and retail process of service delivery (IEA 2005).

The standard model of restructuring based on the UK experience involves functionally separating integrated monopoly utilities and creating markets for different aspects of the electricity sector: generation, transmission, distribution, and retail (see Table 4.1). Victor and Heller (2007, 6) outline the four steps:

1 Generation, transmission, and distribution are unbundled.
2 Parts of the public utility are privatized.
3 Regulatory institutions to oversee conduct in the newly "competitive" parts of the system are created.
4 Power pools for electricity trading are created.

This restructuring takes place in order to open up space for private-sector accumulation through, for example, independent power producers (IPPs), open retail markets, and power trading. The role of the public sector in a restructured system then shifts from an active manager of and participant in all aspects of the sector to a more circumscribed role as market rule–enforcer and referee to national and international corporations.[1] This scale-back of state participation has also been accompanied by a shift more generally from command-and-control regulation of the sector to market-based voluntarism and, in cases like California, *de*regulation (Doern and Gattinger 2003). These reforms are accompanied by the myriad challenges of coordination and policy coherence, given the complex systems that emerge.

Restructuring is underpinned by neoliberal ideology wherein the appropriate governance framework for electricity requires private actors and markets to play the central allocative roles. As an article in *The Economist* (2011) put it, "The rigidity of the public sector does not merely reduce the quality of services. It also discourages innovation. In the private sector innovative firms routinely experiment with new business models, measure the success of those models and then expand successful ones." How this ideological commitment manifests in practice in various jurisdictions differs. In the case of the United Kingdom, challenging the power of the coal unions played an important role so that a shift to new sources of energy became part of restructuring. In California, Enron and the private sector saw an opportunity to open new markets for accumulation. The specific policy packages are not uniform in either speed or scope, but they move power sectors along the continuum from integrated service provision toward market-based private systems.

Around the world, restructuring is justified with reference to a wide range of factors: debt reduction (via sale of public assets), reducing economic inefficiencies (via competition), and most recently, a need to shift to renewables (via innovation). This latter link between innovation, markets, and renewable (or green) power is complex and important; it's addressed further in Chapter 5. Briefly, however, the hope for environmentalists long frustrated with fighting against centralized coal and nuclear systems is that broken-up and restructured systems will allow for a broader diversity of (ideally, new and greener) technologies and actors (Cohen 2006b). In Canada, the power of nuclear and fossil fuel industries is significant, as is the access these groups have to policy makers. So whereas market-based restructuring sets a framework for the type of renewable systems that emerge, so too does the degree of regulatory capture of federal and provincial state agencies (Bratt 2012; Durant 2009; Rowlands 2007).

Renewable energy can be smaller scale, empowering homeowners and community groups to participate (Scheer 2007). Herman Scheer and other advocates of "soft energy paths" advocate for generation sources that are small, appropriate, and distributed near demand (Lovins 1977). The need to move toward more renewable power and increased democratization of the power sector (in terms of policies and actors) has led some environmental groups – the Sierra Club in California and Energy Probe in Ontario, for example – to support power-sector restructuring. However, restructured systems may in some ways undermine the very policy levers and political will required for conservation and widespread renewables development (Cohen 2006b; Heiman 2006). There are also significant problems of efficiency, cost, and implementation of widespread distributed systems (Akorede et al. 2010; Jenkins, Ekanayake, and Strbac 2009; Quezada, Abbad, and San Román 2006).

Three distinct but often co-existing visions of electricity governance have thus emerged out of contemporary restructuring debates: public/state, private, and social economy/eco-local (see Table 4.1). Within each, a different conception of the role of electricity, and prescriptions for how it should be structured, owned, and regulated, exist. The main shift today is between vision one (public/state) and vision two (private), though developments in Germany and particularly in Denmark have aspects of vision three. In both, generation tends to be large scale and geographically concentrated because of cost and resource efficiencies. The key difference is how the benefits and costs of power generation, transmission, and retail are structured. The deep

TABLE 4.1
Electricity governance regimes

	Public/state	Private market based	Social economy/ eco-local ideal
Conception of electricity	Core public good	Commodity	Local asset
Financing/ investment	Public debt	Private investors	Local investors/ users
Generation	Large scale, geographically concentrated in resource-rich areas	Large scale, geographically concentrated in resource-rich areas	Smaller scale, distributed and diversified based on local generation assets
Role of state/ regulation	To ensure public utilities provide maximum public benefit	To coordinate and develop markets; address market failures where necessary	To ensure true cost pricing for environment and privilege local ownership and grid access
Transmission/grid structure	Centralized	Coordinated through central agency to facilitate open access and reliability	Decentralized

green and social economy ideal is far less mainstream than the others and will be explored more fully in Chapters 6 and 7, but it is important to note here that this vision is increasingly being used as a foil for the kinds of developments that support less environmentally or socially empowering forms of restructuring. In reality, what is taking place is a somewhat misleading rhetorical opposition between vision one (public/state) and the other two (private and social economy/eco-local).

Roots of Restructuring: Chile, the United Kingdom, and the United States
Power-sector restructuring first emerged as an issue in Chile, the United Kingdom, and the United States in the late 1970s and then spread via transnational actors over the next three decades to many countries around the

world. In both Chile and the United Kingdom, governments embarked on radical reforms of previously publicly owned power systems. Also, in both cases, ideologically driven actors rather than consumer demand played central roles in reform (Beder 2003). In Chile, the overthrow of the Allende government in 1973 ushered in Pinochet's military dictatorship. The latter, together with an influential group of Chicago school economists – including Milton Friedman – privatized state-owned utilities, unbundled portions of the power sector, and allowed wholesale retail competition with the country's 1982 Electricity Act. In 1986, an independent system operator and open retail markets were created. Chile's power sector is now 100 percent private, and assets are owned by foreign multinationals, including Spain's Endesa and American-controlled AES Gener (Anderson 2009; Hall 1999).

In the United States, restructuring efforts aimed at breaking up private and public monopoly utilities and creating markets for independent (nonutility) generation and power trading. At the federal level, the push to restructure began in 1978 with the Public Utility Regulatory Policies Act, in which Congress mandated that utilities purchase power from private (nonutility) generators with the condition that this power was obtained at lower cost than from incumbent utility sources. According to Heiman and Solomon (2004, 97), the act emerged "in part, to circumvent monopoly control over power provision that was threatening to block the energy independence programs of President Carter." Although Carter was concerned with the OPEC crisis of 1973 and energy security, subsequent administrations dropped the federal support for renewables and energy security but deepened market restructuring via rulings by the Federal Energy Regulatory Commission (Sovacool 2011).

Pressure to restructure, then, was driven primarily by private actors and facilitated by state policy. For example, the three major private power utilities in California spent $69 million between 1994 and 2000 pushing deregulation (Beder 2003, 96). The state passed legislation – AB 1890 – in 1996, which deregulated private utilities, created an independent system operator to manage transmission systems and opened the retail market to competition (Anderson 2009; Burtraw, Palmer, and Heintzelman 2000). A "benefit to consumers" argument played an important role in justifying these moves, as did reference to the new opportunities for nontraditional actors (communities) and generation sources (including wind and solar) (Anderson 2009). In the words of the US Government Accountability Office:

The federal government has pursued a policy to restructure the electricity industry with the goal of increasing competition in wholesale markets and thereby increasing benefits to consumers, including lower electricity prices and access to a wider array of retail services. In particular, federal restructuring has changed how electricity is priced – shifting from prices set by regulators to prices determined by markets; how electricity is supplied – including the addition of new entities that sell electricity; the role of electricity demand – through programs that allow consumers to participate in markets; and how the electricity industry is overseen – in order to ensure consumer protection. (United States Government Accountability Office 2005, 2)

The United Kingdom was the first major industrialized nation to fully restructure its power sector and is considered by some the gold standard in restructured electricity market design (Joskow 2009). Although the Public Utility Regulatory Policies Act set the framework for restructuring in the United States, that country's federal system meant that the degree of restructuring often varied from state to state. The 1983 Energy Act and 1989 Electricity Act under Margaret Thatcher's Conservative government went on to set what is now known as the standard model of power-sector reform. Public electric utilities weren't privatized until 1990. These initiatives were part of a broader package of neoliberal restructuring in that country throughout the 1980s and 1990s. They also served to undermine the power of the British coal unions and the labour movement in the country. This period saw the sell-off of British Telecom, British Steel, British Petroleum, British Airways, and a wide range of other companies restructure governance in the country (Beder 2003, 198–99).

From these first key experiences, a powerful policy package and supportive narrative emerged, promoting the virtues and customer benefits of restructuring power systems. These actors and forces continue to push for restructuring today through a network of think tanks and government agencies across the developed and developing worlds. According to Gratwick and Eberhard (2008), actors in the World Bank, International Monetary Fund, London Economics (a consulting firm), and Oxera, as well as think tanks such as the Adam Smith Institute and the Heritage, Cato, and Fraser Institutes, played key roles. Gratwick and Eberhard argue (2008, 3949): "A number of the consultants involved in the reforms in Chile, Argentina (later) and the United Kingdom, subsequently were involved as advisers to development finance institutions and developing country governments, and were often directly involved in the design of power-sector reform in developing countries." Countries that have subsequently reformed their power sectors include India, Tanzania, and South Africa. Among industrialized nations, New

Zealand, Australia, and, in Canada, the provinces of Ontario and Alberta, have also significantly restructured their power sectors, implementing many if not all of the standard model of reforms.

Drawbacks of Electricity Restructuring
Power-sector reforms have been and continue to be problematic in the jurisdictions where they've occurred. These reforms include price manipulation in power markets by powerful actors (Enron, for example), the coordination costs of a decentralized system, and the environmental costs of privileging markets when environmental degradation and climate change loom large. Both technical and political economy challenges emerge from these deeper debates around the design of power systems. When grids need upgrading, who pays? When they have limited capacity, which actors are allocated priority? How much extra capacity is needed for system reliability, and how much for demand growth that can potentially be mitigated?

Technical
The specific technical requirements of generating and transporting electricity make restructuring the sector a particular challenge. Two technical limitations – capacity and reliability – make central coordination of the flows through the system a vital part of electricity infrastructure. This need for systemic coordination, together with the sector's importance to society, has necessitated significant public investment and oversight in the electricity sector and, until fairly recently, heavy regulation (Bouffard and Kirschen 2008; Egan and Turk 2008; Freitas et al. 2007). Electricity system capacity represents the ability of the power system to handle overall increased power generation.[2] This is a challenge because if a generator connects to a power line already at capacity, it can result in an oversupply, and all parts of the grid are put at risk. In 2003, perhaps the largest blackout in history resulted in tens of millions of people being without power across Ontario and the northeastern and midwestern United States. Reliability is a function of a highly interconnected power system wherein the system is more stable because shortages in one region can be supported by increasing power from another (balancing out the system). Where variable power sources like wind and solar make up a significant part of the mix, reliability is a particularly important issue. This is significant because it encourages integration within a power system, which then requires central coordination to deal with capacity and load issues.

As a result of these two technical issues, a key challenge emerging from restructured systems is that restructured markets don't result in genuine competition and always require significant government support – via investment, coordination, and regulation (Anderson 2009). Secure and uninterrupted supply of electricity is, in a broad sense, a public good (Abbott 2001; Nelson 1997) with monopoly characteristics.[3] The finished product is unlike those in many economic sectors, as it is difficult to store after it is produced in large quantities.[4] This means that, at all times, demand and supply (voltage and current) of the power on the lines needs to be precisely matched. Electricity transmission and distribution requires managing, and sometimes prioritizing user demand within available infrastructure is important (Abbott 2001; Nelson 1997; Thomas and Hall 2006).

Market restructuring increases administrative coordination required in this complex sector – and thus risks of expanded grid interconnection – between actors with significantly different goals and incentive structures. For example, the old public utility often retains a degree of market power by virtue of its size and, prior to reform, its management of grid access. In restructured hybrid systems, independent system operators are sometimes created to replace utilities to manage the grid and ensure open access for newly competing generators (i.e., IPPs), and to engage in transmission planning and oversight. However, Lyster (2005, 419) points out that "designing such an institution does pose significant difficulties for matching risks, responsibilities, and authority." Interconnected systems enhance reliability with a high degree of regulation, coordination, and system management. Adequately managing these systems across national and provincial borders raises complex issues of regulatory power. It also raises distributional issues about the allocation of costs and benefits. This becomes particularly problematic when public needs require power systems to be reliable, affordable, *and* environmentally sustainable. More interconnections can mean more instability, as evidenced in the rolling blackouts across Ontario and the northeastern and midwestern United States in 2003 (NEB 2009).

Political Economy

In addition to the technical challenges of managing a functionally separated market system, restructuring allows private actors to engage in profit-seeking activities that can damage public interests. California's spectacular failure in restructuring in 2000–01, for example, illustrates the public's vulnerability to market manipulation by energy companies (Beder 2003; Cohen 2004).

In its electricity crisis, rolling blackouts and price spikes of 800 percent resulted from lack of regulation and profit-maximizing activities of Enron, the now infamous company that manipulated power shortages for private gain. Following this debacle, the enthusiasm for rapid power-sector deregulation in North America has dimmed but not abated (IEA 2005).

The claims of reform advocates about market efficiencies, competition, and renewable energy are challenged on several fronts (Abbott 2001; Anderson 2009; Dubash and Williams 2006). What has emerged from empirical research is a set of cautionary lessons detailing how the promised benefits of power-sector reform rarely occur in practice. Electricity consumers are a captured market, and public bodies are under pressure to ensure reliable and reasonable access, whether they own generation or not. The partially reformed systems that have emerged let the public sector assume these costs while the private sector takes the profits from generation and trading yet doesn't have to take on the systemic risk or management (Cohen 2006b). Private actors in restructured markets thus have incentives to cost-shift unprofitable infrastructural investments to public agencies.[5] The short-term nature of private actors creates incentive problems for long-term sustainability, as they tend to discount important long-term investments that are essential to rational environmental calculations (Barkin 2006; Mitchell 2008). For example, power grids are expensive to maintain and are required (particularly in Canada) to stretch over long and often sparsely populated distances (Achenbach 2010). The distance factor negatively impacts the willingness of private for-profit firms to build universal access, and means that public funds are nearly always required in the industry. This raises issues of inadequate system maintenance and investment in upgrading infrastructure to meet new demands. Neoliberal restructuring is at odds in many ways with the interventionist state needed to address not only coming environmental challenges but fuel-poverty ones as well.

As a result of these issues, many analysts have found the outcomes of power-sector restructuring wanting. For example, John Anderson (2009, 82–84), president of the Electricity Consumers Resource Council in the United States, argues that blind ideological faith is not a useful route to the creation of functioning markets, and he derives lessons from the practice (rather than theory) of reforms around the world. He argues that the benefits have accrued to suppliers; the drawbacks, to the average power consumer – a position echoed by some Organisation for Economic Co-operation and Development commentators (IEA 2005; OECD 2001). Restructuring has also not led to "free" or even particularly competitive markets, as powerful (often multinational) corporations dominate. Adequate regulation and system coordination is plagued by informational asymmetries between the actors in restructured systems. Anderson (2009, 84) concludes by saying, "It is striking that industrial and other consumers from around the world have come to the conclusion that today's 'restructured' markets are far from the competitive markets that they envisioned and that these markets have failed to achieve the stated goals and provide net benefits to consumers." Taken together, these issues and critiques of restructuring illustrate how continued and significant public involvement in the power sector is both necessary and desirable.

North America: Toward a Continental Power Market

Power-sector restructuring internationally exerts pressure on Canadian provincial systems. Actors in the United States such as the Federal Energy Regulatory Commission (FERC) and the North American Electric Reliability Corporation (NERC) work as part of a neoliberal tag team with domestic agencies and actors, advocating reform in Canada. These efforts continue, despite the problematic results of restructuring in reformed systems, in part because of the relative wealth of hydroelectric resources in Canada and the desire to export to US markets. Regulatory leadership from the United States and private gain, and not reliability and environmental needs, motivate the policy shifts described in the following chapter. One aspect of these shifts is that continental grids are being strengthened (Gattinger and Hale 2010). The opening up of provincial grids to private entities, and strengthened north-south transmission lines, provides opportunity for power for financial gain from electricity market trades. Given the environmental consequences of expanding generation and consumption, this trend is problematic. I illustrate these changes in three parts: (1) the export orientation of some provinces (British Columbia, Manitoba, Ontario, and Quebec), (2) the focus of new transmission for north-south export, rather than reliability, and (3) US regulatory pressures (via FERC and NERC) pushing harmonized market structures.

The quest to integrate more fully with US electricity markets is represented across Canada in the publications of the National Energy Board (NEB 2009, 2010b), Natural Resources Canada (2008), provincial energy ministries (Ontario 2011), and industry associations such as the Canadian Electricity Association (Egan and Turk 2008). Roger Goodman at the Canadian International Council (2010) argues that there needs to be *more* integration at the policy level so that Canadian resources and production potential can be exploited to its fullest, a position popular in the oil and gas sector. Goodman's position is echoed by governments and businesses focused on selling to the vast market south of the border (de Villemeur and Pineau 2010; Guimond 2010).

Canada-US Electricity Trade
Canada and the United States have one of the largest trading relationships in the world, and this close interconnection holds true for the power sector as well. This means that changes in the United States have significant (and disproportionate) impacts on Canadian exporters and, by extension, Canadian citizens. Power exports play an increasingly important role across Canada. Exports of electricity to the United States grew by 215 percent from 1990 to 2008. Since 2008, net electricity exports have continued to grow: from 29.6 net terawatt hours (TWh) to 52 net TWh in 2012 (NEB 2008, 2012). These 52 TWh of net exports represent almost 10 percent of the total power generated in Canada in 2012. As trade links increase, they create continued pressure on provincial electricity governance and infrastructure, particularly as market reform has become a condition for access to US export markets (Cohen 2004, 2).

International electricity exports help explain why Quebec and British Columbia, for instance, are under pressure to allow private actors to generate and sell power. The electricity sector is a target for reform in part because generation of power in Canada is big business, netting $2.1 billion for Canadian exporters in 2013 (NEB 2014). Although exports vary from year to year, the largest Canadian exporters in 2012 were Quebec (24,037 GWh), Ontario (13,823 GWh), British Columbia (10,839 GWh), and Manitoba (8,048 GWh). Table 4.2 illustrates the differences between provinces exporting power to the United States in 2012, including the total export revenues and the proportion of generation exported. Manitoba, for example, exported 24 percent of its total generation that year. The prevalence of hydroelectric power storage in British Columbia, Manitoba, and Quebec allows for electricity storage (or banking) that other provinces do not enjoy to the same degree.[6]

Electricity export growth is predicted to only increase in the future as United States policy shifts toward enhanced energy security and clean power development (Goodman 2010). The United States is currently under pressure to meet environmental policy goals, since demand is increasing, and the country is heavily dependent on coal for almost 50 percent of its electricity.

TABLE 4.2

International electricity exports by province, 2012

	MWh	Revenue (million $)	Main exporting companies (MWh); (total number of exporting companies)	Price/ MWh ($)	Exports as % of provincial generation[1]
QC	24,037,874	894	Hydro-Québec, Brookfield, MEHQ; (6)	33.27	13
ON	13,823,309	425	MEHQ, OPGI, Powerex, MAG Energy, Royal Bank; (29)	30.24	10
BC	10,838,969	274	Powerex, BC Hydro; (2)	25.33	18
MB	8,048,290	267	Manitoba Hydro; (1)	33.16	24
NB	778,732	51	NB Power, Twin Rivers PCI, Algonquin Tinke; (5)	65.93	7
AB	41,582	2.2	TransAlta, CP Energy Marketing, TransCanada Energy; (5)	53.94	0.1
SK	68,506	2.1	Northpoint, Rainbow; (2)	31.49	0.3
NS	1,094	0.04	NS Power; (1)	44.46	0

1 Exports are complex and include more than provincial generation. They are also a function of actors in one province purchasing power (e.g., British Columbia with Alberta) from neighbouring jurisdictions and selling it. *Source:* NEB 2013, 2012; Statistics Canada 2013b.

As Canadian provinces restructure electricity markets to allow for private ownership and competition, they become subject to international trade rules governing services (Cottier et al. 2010; Wilke 2011; WTO 2011, 2013b). When opened to private-sector actors, industries fall under the North American Free Trade Agreement (NAFTA) and General Agreement on Trade in Services (GATS) rules regarding market access and competition (Cohen 2004) and are locked into a supranational conditioning framework (McBride 2005). Under the national treatment clause of NAFTA, for example, Canadian provinces are required to treat private US firms just as they treat Canadian firms in the same area. That said, it remains legitimate for the provinces to maintain integrated monopolies in the public interest. Provinces that open access for private retailers, generators, and wholesalers in order to gain access to US export markets are thus paving the way for significant limitations on what governments can (and cannot) do to direct the price and structure of electricity in the future domestically. Consequently, initiatives to deregulate, reregulate, and restructure electricity in the United States have significant spillover effects in Canada (Cohen 2004, 2006a; Horlick and Schuchhardt 2002).

Regulatory Harmonization: FERC and NERC

Access to US markets is vital for export-oriented provinces; this access is regulated by FERC in the United States. More interconnection with the United States and subsequent policy harmonization with FERC mean that Canadian actors increasingly need to conform to US regulatory standards. J.O. Saunders argues that

> FERC, through its rulings on reciprocity of access, has had an important influence on the structuring of the Canadian electricity industry ... Moreover, it has exercised this influence while all the time insisting that it was deferring to the jurisdiction of Canadian regulators in Canada. What the FERC was essentially doing in its rulings, however, was overruling the principle of national treatment set out in the NAFTA (and the GATT) and replacing it with the principle of reciprocity. Put differently, it was replacing the principle of *free trade* with the principle of *fair trade*, with FERC as the adjudicator of what is fair. (Saunders 2001, 171; emphasis in original)

Since 2007, the North American Electric Reliability Corporation (NERC) has enforced electricity reliability standards in the United States and also in Ontario and New Brunswick. It is seeking to expand this to include other Canadian provinces and Mexico. Prior to 2005, NERC was the national regulatory body for American electric reliability standards. However, after a working group in 2004 found that the 2003 blackout affecting 50 million people was preventable, and was due to the failure of actors to comply with voluntary reliability standards, pressure developed for mandatory binational standards. The NERC applied to FERC for a name change after the introduction of the 2005 Energy Policy Act; it became the North American Electric Reliability Corporation (formerly Council). Even though Canadian agencies have a seat at the table, the United States has greater influence and US agencies approve standards before Canadian ones (Gattinger 2010). NERC and FERC have been endowed with the power to enforce financial and export penalties to more effectively ensure compliance.

Canada's national energy regulator, the National Energy Board (NEB), has willingly ceded its regulatory role. It announced in 2010 acceptance of NERC standards and said that it was working with provinces to implement them:

The NEB has recognized the North American Electric Reliability Corporation (NERC) as the Electric Reliability Organization in North America, as
applicable to IPLs. In 2007, NERC reliability standards became mandatory
in the United States. Canadian regulators, including the NEB, are working toward the implementation of mandatory standards in their respective
jurisdictions ... For instance, NERC standards are adopted through legislation in British Columbia and Alberta, and are mandatory in Ontario and
New Brunswick through the market rules governing transmission in those
provinces. NERC standards apply in Saskatchewan and Manitoba through
contractual agreements with the Midwest Reliability Organization. (NEB
2010b, 31)

The abdication of Canadian federal responsibility for managing electricity
systems in the national interest means that FERC is left as the de facto regulator for the continent.

FERC regulatory changes push the unbundling and marketization of
electricity here in Canada. For example, in 1996, FERC issued order no. 888,
which required utilities that operated and owned transmission to file open
access transmission tariffs (OATTs). The stated goal was to facilitate competition and bring more efficient, lower-cost power to Americans (Cohen
2004). Canadian provincial governments see their interests in conforming
to FERC requirements for both ideological and export-driven reasons.
FERC requirements are not simply about system reliability and have led directly to market restructuring in Canadian provinces. Without restructuring,
for example, Canadian exporters were denied export permits to the United
States.

Ian Blue, Q.C. (2009) covers the important history of the relationship
between FERC and provincial electricity bodies, illustrating that in 1995 a
Hydro-Québec partially owned company (Energy Alliance Partnership) was
denied access to sell at market rates into the United States by FERC on the
basis that Hydro-Québec had "too much market power." In 1997, FERC
rejected (BC Hydro–owned) Powerex Corporation's application to sell electricity because of BC Hydro's market power. It also rejected Ontario Hydro's
export application in 1997 based on its market power. In contrast, in 1996,
FERC approved TransAlta's application because of Alberta's 1995 market
restructuring (Blue 2009, 344–45).

These decisions led to a clear set of policy changes in exporting provinces. Once British Columbia and Quebec filed OATTs in 1997, FERC
approved their export permits. Ontario started reorganizing its electricity

system in 1998 and, like the others, was then allowed to sell power to the United States. In Nova Scotia, one policy analyst at the Department of Energy put the drive to market reforms this way:

> This was being driven at the time by a lot of influence from FERC; for example, any of the utilities interconnected with the United States were being forced to move to FERC proforma open access transmission tariffs and ... to open their wholesale markets to competition to access US markets.

> So because we [Nova Scotia] wanted to have reciprocity with our New Brunswick neighbour – our main system operator for the Maritime region here – we moved forward after the electricity marketplace governance committee report made its recommendation with essentially creating or forcing Nova Scotia Power to create a FERC proforma open access transmission tariff. We opened up our wholesale market, which is six municipal electric utilities, to competition, so they're now open for competitive supply. And we've established an initial set of market rules under the Electricity Act. (Personal interview, May 19, 2010)

Interprovincial competition for export ability also plays a role in market restructuring. Just as governments use FERC rules as justifications for power-sector reform, private actors can apply to FERC for rulings against integrated public utilities. The province of Alberta lodged complaints with FERC against BC Hydro, alleging that the utility was not granting it nondiscriminatory access to move power through the province (Alberta does not have direct access to the Western Grid). In 2003, the BC government separated the transmission components of Hydro into a separate company, the BC Transmission Corporation (BCTC). This was done to conform to FERC preferences for separation of functions in order to provide nondiscriminatory access for private entities to the transmission system. One estimate put the cost of this separation at $65 million (McMartin 2010). In 2010, British Columbia's Clean Energy Act reversed this separation and again rehoused transmission within BC Hydro in order to increase efficiencies in planning and coordination.

Strengthening the Continental Grid

The continental push for electricity systems in Canada is represented in the concrete north-south orientation of transmission infrastructure. Increasing the number of FERC- and NERC-regulated interconnections strengthens the pressure on Canadian provinces to continue restructuring. International transmission construction has intensified since the signing of NAFTA. There are nine new lines to the United States, to be completed in the next five years, further strengthening the export capacity of Canadian provinces. Appendix 5 contains a list from the National Energy Board of major proposed international power lines (IPLs). According to an NEB presentation in 2011 to the New England–Canada business council, there are more than thirty major transmission interconnections between the two countries (NEB 2011). The most recent, a 1,000 MW IPL, connected New Brunswick and Maine and doubled the interconnection between the two jurisdictions.

These continental changes impose significant infrastructural costs on provinces and, ultimately, on ratepayers. Private actors, be they energy traders or generators, seek to maximize profits by directing electricity toward the consumers paying the highest rates and externalizing management costs and risks to the public sector. The buy-low, sell-high mentality that forms the modus operandi of private markets results in large volumes of power being moved from one jurisdiction to another. According to the NEB:

> Electricity transmission over IPLs has almost doubled since electricity markets started to restructure in the mid-1990s. Imports from the US have increased as demand growth has outpaced supply growth in provinces like Ontario, BC and Alberta. The north-south trade exploits the complementary seasonal peaks between the winter heating demand in Canadian provinces and the summer cooling demand in American states. (NEB 2010a, 29)

Export for private profits for generators and private grid owners, rather than reliability, is an important driver of publicly funded grid expansions. In Alberta, new transmission lines have been particularly contentious: critics charged they were being built for exporting electricity to the United States rather than for local need or reliability. The 345-kilometre Montana-Alberta Tie-Line, for example, owned by Enbridge, connects Alberta to the Western Interconnection as of 2013. The Alberta government maintained that reliability, rather than export, was the goal and passed Bill 50 in 2009, removing the requirement for needs assessment hearings on the $16 billion transmission system investments. WikiLeaks cables between US and Alberta politicians from 2003 and 2008, released in 2011, "show that Alberta politicians offered to export power to the United States using excess electricity generated by oil sands facilities" (Nikiforuk 2011). Oil sands developers have subsequently reduced their estimated cogeneration capacity and are now expected to be net consumers (rather than producers) of power (Genalta Power 2011).

The International Energy Agency estimates that US$7.6 billion per year of electricity infrastructure investment will be needed in Canada from 2005 to 2030. The NEB cites the positive role of interconnections for increased reliability within the system (NEB 2009). It argues that "this inter-jurisdictional trade provides reliability benefits and increases overall system efficiency" (2010b, 29). Since electricity needs to be used and is difficult to store (hydro being an exception), grid interconnections with other provinces and US states provide stability within the system, protecting against power shortages or overloads, and facilitate export. Of course, these reliability benefits can also be derived from national, rather than international, interconnections, as well as through the development of district heating and CHP capability. These tradeoffs become even more important going forward, as pressure for new renewables, together with distributed generation and smartgrids, become a larger focus for provinces.

Summary

Power-sector restructuring – starting with Pinochet in Chile and Thatcher in Britain – was ideologically driven, rather than necessary for efficiency or consumer choice, as often claimed. In many countries, the United Kingdom excepted, the ultimate result was not the creation of competitive markets and lower prices, or the promised environmental gains spurred by the private sector. Indeed, the results have been disappointing on most fronts except for those corporate actors that gained new avenues for profitability. Where shifts away from coal-based generation have occurred, it has been as a result of public funding and public policy, working together with strong social movements, as the next chapter illustrates.

Far from shifting toward the social economy/eco-local governance model, then, the international forces shaping Canadian electricity emphasize increasing generation, not sufficiency, and less rather than more democratic control. Expanded continental electricity markets are designed to give Canadian exporters access to US markets, and to open Canadian markets to private investment. These market structures for retail and wholesale then put pressures on Canadian transmission infrastructure by requiring support for larger volumes to be traded over ever-increasing distances. The rules of the expanded grid and export access are increasingly shaped by the electricity market and by governance structures in the United States via FERC and NERC, as well as expanding international trade in services agreements. These trends matter a great deal for co-operative development in the power sector, as community-based actors are but a small part of these broader developments since the 1970s.

5

Continental, Private, and Green(er)?

Canadian Electricity Restructuring

Electricity ownership across Canada – particularly for new renewable-generation projects – is shifting to the private sector. Domestic policies, together with the international forces outlined in Chapter 4, are restructuring provincial power regimes away from the post–Second World War public systems and toward deeper marketization. These developments have implications not only for the potential of electricity co-operatives but also for the reduction of future carbon emissions and the enhancement of energy security. In this chapter, I focus on the interplay between electricity restructuring policies, renewable generation, and sustainability in Canada. I argue that electricity reforms taking place in Canada are not uniform; they are being implemented in provincially distinct and often piecemeal ways. Three major changes stand out: more private power generation, deepening continental power grids and markets, and more alternative renewable generation. Canadian reforms rarely include wholesale utility privatization (as happened with Nova Scotia Power in 1992); rather, they involve incremental market restructuring justified with reference to environmental sustainability, demand growth, and new renewables development.

Although they open space for electricity co-operatives, these changes are often problematic for Canadian communities. This is particularly the case when private ownership of new renewable generation is subsidized by the public sector. Canadian reforms, like those in other countries, emerged not out of economic necessity or even commitment to environmental principles but in order to create markets and avenues for private accumulation in sectors previously considered off-limits.[1] For example, a significant increase in IPPs has occurred in many provinces. With this entry of private power generators has come a call from US companies to gain equal access to Canadian power markets, long protected by the public utility structure. This restructuring push is thus accompanied by significant long-term social, financial, and environmental implications for Canadians.

Despite these developments and pressures, most provinces in 2016 retain significant public control and ownership of generation assets. Restructuring policies have not yet succeeded in privatizing power to the degree that exists in the United States, making critical analysis of current policy initiatives even more pressing. Coal generation–based provinces (Alberta and Nova Scotia) have an environmental need to diversify generation sources. Yet, it does not necessarily follow that diversification of sources need be rooted in private accumulation and increasing generation. Provincial moves to green power generation are being used to justify further restructuring of public power systems, even though renewable hydroelectricity is the dominant fuel in several provinces (including British Columbia, Manitoba, Newfoundland and Labrador, and Quebec). This is accomplished in part via a definition of "green" that excludes both large-scale hydro and nuclear generation (Weis et al. 2009). This information helps explain how, where, and why co-operatives find themselves with the opportunity and the need to participate in provincial power sectors.

Canadian Power: (Relatively) Public, Cheap, and Renewable

The data on Canadian electricity by source, generation, and price reveals important differences from our southern neighbour, differences that, in many cases, undermine the rationales for market reforms. The Canadian power sector is characterized by a far greater degree of public ownership than that of our major trading neighbour, the United States. The dominance of public generation sources – 72 percent of installed capacity in 2011 – and the relatively low prices in hydro jurisdictions mean that, on the whole, electricity in Canada is public, cheap, and (again, relative to our southern neighbour) renewable. Provincial differentiation falls along two axes: share of renewable generation (high/low) and degree of private participation (high/low) (see Table 5.1). This is important, because it demonstrates that, in Canada, renewable power was developed by the public (not private) sector because of its willingness to take on long-term debt and build infrastructure for public need rather than private profit.

TABLE 5.1
Majority ownership and fuel source by province, 2012

	Renewable	Fossil fuel
Public	BC, YT, MB, QC, ON, NL	SK, NB, NU
Private	PEI[1]	AB, NT, NS

1 Prince Edward Island imports the vast majority of its electricity from New Brunswick. So, while the renewable installed capacity on the island is a significant percentage (56 percent), the power residents consume is from New Brunswick.

The model for power development in Canada has been, until recently, one of public, vertically integrated utilities. In all provinces save Alberta, Nova Scotia, and Prince Edward Island, Crown corporations are the major source of generation and system management. Provincial electric utilities are vertically integrated, with transmission, distribution, generation, and retail operations owned by the province. In Ontario and Alberta, local distribution companies (often municipal) also play a role.[2] This model of sectoral organization was developed to ensure reliability, availability, and affordability for a given jurisdiction's citizenry. The private sector was, before the public sector stepped in, either unwilling or unable to meet these goals.

Public-sector willingness to take on long-term debt mattered for the development of significant renewable (63 percent hydro) power capacity in Canada. Public investment, together with the available natural resources, played a central role in determining which energy sources were developed in Canada. This is because development of large-scale renewables, like hydro, tends to have significant up-front costs and long payback time frames. Public investment continues to play a critical role in renewables development across the country, albeit without the financial returns of actually owning the new generation in most cases. In provinces with hydro resources, such as British Columbia, Manitoba, and Quebec, investment decades ago is now paying off in the form of lower prices from old generation assets (see the prices for "Hydro" in Table 5.2). Construction costs, when amortized over twenty years, mean that hydro is one of the most cost-effective and (relatively) environmentally friendly sources of electricity.

Private actors were (and still are) unlikely to undertake new renewable-generation projects without significant financial guarantees. For private actors drawing financing from international financial markets, twenty years is too long for a shareholder return (Froschauer 1999). Public power can be distributed on a cost-of-production basis, whereas private-sector generators are required to maximize profits. It is, therefore, not a coincidence that the

TABLE 5.2
Residential electricity price and fuel source, 2015

	2015 ¢/kWh	Main generation source
Halifax	16.03	Coal
Charlottetown	15.62	Wind/imported power (NB)
Ottawa	14.86	Nuclear
Regina	14.37	Coal
Toronto	14.31	Nuclear
Moncton	12.30	Coal
St. John's	11.55	Hydro
Edmonton	11.55	Coal
Vancouver	10.29	Hydro
Winnipeg	8.11	Hydro
Montreal	7.19	Hydro

Source: Hydro-Québec 2015; Statistics Canada 2013a.

relatively restrictive nature of the markets in most Canadian provinces correlates with some of the lowest electricity prices in North America. In 2015, for example, electricity prices in New York and San Francisco were 28.90 and 27.69 cents per kilowatt hour (kWh) respectively, while in Montreal it was 7.19 cents per kWh (Hydro-Québec 2015).

Canada has the fourth-largest national share of hydroelectric generation in the world at 62 percent of installed capacity in 2012.[3] It also has the third-highest total hydropower generation at 376 gigawatt hours (GWh) (after China and Brazil) (IEA 2012b; Statistics Canada 2013b). Hydroelectricity as a generation source is renewable and significantly greener in terms of life-cycle air emissions than coal, natural gas, and diesel and is, over the long term, cost-effective[4] (Hydro-Québec 2003). Nuclear power accounts for 14.5 percent of the share in generation, and steam plants fired mostly by coal make up a further 15 percent of the mix (see Table 5.3). New renewables like wind and tidal power play a small, albeit growing, role in Canada's generation mix. Between 2000 and 2013, wind, tidal, solar, and hydropower grew from 61 percent to 65 percent of total electricity generation, with new renewables like wind, solar, and tidal power growing from 0 percent of the total to 1.52 percent (Statistics Canada 2002, 2014).

Each fuel source – from hydro, wind, and solar to uranium, coal, and natural gas – comes with a unique cost, reliability, and environmental footprint. The vast majority of power generated in Canada comes from large, centralized power plants (between 100 and 5,000 MW capacity).[5]

TABLE 5.3
Canadian electricity generation by source, 2012

	TWh	% share
Hydro	376	63
Nuclear	91	15.3
Conventional steam (coal)	89.4	15
Combustion	29.2	4.9
Wind	8.7	1.5
Solar	0.3	0
Tidal	0.03	0
Total	594.9	

Note: Numbers may not sum to total or 100 because of rounding.
Source: Statistics Canada 2013b.

TABLE 5.4
Life-cycle assessment of GHG emissions (kt CO_2 eq./TWh)

	Minimum	Maximum	CCS min/max
Coal	675	1,689	98/396
Oil	510	1,170	
Natural gas	290	930	65/245
Nuclear	1	220	
Wind	2	81	
Ocean	2	23	
Hydropower	0	43	
Geothermal	6	79	
Solar photovoltaics	5	217	
Solar cross-sector partnerships	7	89	
Biopower	−633	75	−1,368/−594

Source: IPCC 2011, 982.

The changing generation sources of power matter, both economically and environmentally. The technologies used to generate power from these sources vary (seasonally and daily) in their ability to provide predictable fuel on demand (baseload power). Hydro, nuclear, and coal are highly reliable, whereas wind and solar are variable. Finally, the lead time and capital investment needed to develop new generation varies significantly between power sources. Nuclear plants take the longest to develop and are prone to significant cost overruns (Sovacool 2010). Table 5.4 illustrates a recent life-cycle comparison of different generation technologies.

Environmentalist and First Nations groups have articulated many valid and well-documented critiques of the environmental and democratic record of the centralized public utility power model (Brooks 2006; Cohen 2006b; Froschauer 1999; Netherton 2007). What is important today, however, is that the vast differences that exist between different provinces and market structures mean that, in many cases, environmental sustainability may be best advanced through conservation and efficiency measures rather than by creating private power markets for new renewable generation. Within this context, the potential role for co-operatives in either legitimating these new markets or challenging them becomes important for their potential for empowered participatory governance.

Provincial Variation

This federal profile of electricity generation in Canada obscures important differences between the provinces. The provincial structure of the electricity sector means that power-sector reforms are taking place to different degrees across each of the thirteen provinces and territories. Provincial generation-source diversity creates uneven environmental impacts of generation across the country, and with this comes the need for provincial coordination (e.g., for reliability) and targeted policy. Table 5.5 illustrates the concentration of generating fuels by province. In 2012, the provinces break down into

TABLE 5.5
Electricity generation by province and source, 2012

Province	Total TWh	Coal (%)	Oil (%)	Natural gas (%)	Hydro (%)	Wind (%)	Tidal (%)	Nuclear (%)	Other fuel (%)
QC	198.9	-	-	-	96.5	0.5	-	2.1	0.7
ON	154.4	3.1	-	16.1	22	2.6	-	55	2
BC	73.8	-	0.2	3	88.5	0.2	-	-	8.1
AB	66.1	56	0.3	30.5	4	3.5	-	-	5.8
NL	43.7	-	2.5	0.5	96.9	0.2	-	-	-
MB	33.2	-	-	-	97	2.6	-	-	-
SK	21.2	56.5	-	20.1	19.8	3	-	-	2
NS	11.1	49	12.2	20.4	7.7	7.2	0.2	-	3.2
NB	10.2	20	16.1	19.1	27.8	7.1	-	4	5.5
NWT/YT/NU	1.1	-	41	1	57	-	-	-	-
PEI	0.48	-	3	-	-	97	-	-	-

Note: Fuel source may be used, but less than 0.2%. Numbers may not sum to 100 because of rounding.
Source: Statistics Canada 2013b, 2014.

three main groupings: (1) hydro dominant: Quebec, British Columbia, Newfoundland and Labrador, Manitoba, and Yukon; (2) fossil fuel dominant: Alberta, Saskatchewan, Nova Scotia, Nunavut, and the Northwest Territories; and (3) mixed: Ontario, New Brunswick, and Prince Edward Island.

Three provinces stand out in their consumption of fossil fuels for electricity generation: Alberta, Saskatchewan, and Ontario. Overall, in 2012, Alberta used 59 percent of all coal used for electricity generation in Canada in that year (19 million metric tonnes) and 26 percent of all the natural gas, whereas Ontario consumed 5.6 percent and 48 percent respectively and Saskatchewan consumed 27 percent and 12 percent respectively (Statistics Canada 2014). This diversity leads to an uneven environmental impact of electricity generation across the provinces, ranging from 930 kilograms of CO_2 per megawatt hour (kg CO_2/MWh) in Alberta, to just 6 in Quebec (Bell and Weis 2009, 7).

The share of public and private ownership of generation between the provinces also varies. As Table 5.6 illustrates, Nova Scotia is the only province where the ownership of generation is almost wholly in private hands. In all other jurisdictions save Nunavut, which is entirely in the public sector, there is some combination of public and private generation. In the Northwest Territories, for example, industry-owned generation accounts for 51 percent of the total.

TABLE 5.6

Federal and provincial installed capacity by ownership, 2011

	Public (%)	Private (%)	Industry (%)	Total (GWh)
AB	15	**69**	16	12,622
BC	**75**	9	16	15,136
MB	**98**	2	0	5,659
NB	**83**	14	3	3,950
NL	**94**	4	2	7,419
NS	0	**98**	2	2,683
NU	**100**	0	0	54
NWT	48	1	**51**	183
ON	**65.6**	32.2	2.2	36,160
PEI	18	**81**	0	280
QC	**87.5**	4.5	8	41,510
SK	**87**	12	1	4,204
YT	**93**	7	0	128
Canada	**71.6**	21.5	6.5	129,991

Note: Dominant ownership sources are indicated in bold.
Source: Statistics Canada 2013a.

Provincial Policy Shifts: Private Markets and Renewable Generation

It is within the framework of austerity that space is opened up to justify belt-tightening measures that amount to a transfer of assets – and thus power – away from public control. In the power sector, one form this takes is outright privatization, as in the case of Nova Scotia Power. In 1992, the public provincial utility was sold to private investors in what was then the "largest private equity transaction in Canadian history" (Halifax Media Co-op 2010). The justification given for the sale of the utility was to raise capital to service the provincial debt. Avoidance or reduction of public-sector debt plays an important role in the neoliberal rhetorical tool kit in some Canadian provinces. In Ontario, the justification for restructuring was the high cost of new (nuclear) generation and the debts of Ontario Hydro. In this section, I outline how and where provincial power sectors have been restructured across the country, with particular focus on Alberta, Ontario, and British Columbia. These changes have been damaging to both public pocketbooks and public control.

Participation by for-profit actors is increasing in the sector across the country thanks to strong regulatory and financial support from provincial governments, from British Columbia to Nova Scotia (Datamonitor 2010; Nova Scotia Department of Energy 2010; OSEA 2009). These began in the late 1990s. Table 5.7 illustrates the initial restructuring policies by province. These include changes to retail, transmission, and generation of power in nearly all Canadian provinces. For example, the signing of open access transmission tariffs (OATTs), which set rates and rules for actors to move power over provincial power grids, facilitates wholesale (and in some cases retail) power trading. Other important policy initiatives involve the unbundling of integrated utilities (as in British Columbia and Ontario), the establishment of independent system operators, the creation of power pools (as in Alberta), and policy initiatives mandating that public utilities purchase power from private generators. These initiatives began in the mid-1990s and are continuing today. The most recent policy initiatives provide financial incentives for private renewable electricity through, for example, feed-in tariffs.

As a result of these provincial policy initiatives, markets to buy, sell, and trade power either openly or with the public utility have created more incentives for profit and for private actors in the electricity sectors across the country. From 2000 to 2011, private generators increased their share of installed generating capacity from 11.5 percent of the Canadian total to

TABLE 5.7

Initial provincial electricity restructuring policies

Province	Policy	Details
Alberta	1996 Electric Utilities Act (EUA)	○ Created power pool ○ Opened transmission (OATT)
Ontario	1998 Energy Competition Act	○ Unbundling of transmission, distribution, and generation (integrated utilities) ○ Breakup of Ontario Hydro ○ Creation of wholesale and retail electricity markets (opened May 2002)
	2004 Electricity Restructuring Act	○ Tasked public agencies with incentivizing the development of new generation ○ Started making large IPP power calls for new renewable generation
	2009 Green Energy and Green Economy Act	○ Brought in North America's first feed-in tariff (FIT) (guaranteed price contract) for renewable generation
Quebec	1997 filed OATT	○ Opened transmission grid to private generation
	2006 Energy Policy	○ Ended moratorium on private hydropower below 50 MW ○ Started making large IPP power calls for new renewable generation (wind in particular)
British Columbia	1997 filed OATT	○ Opened transmission grid to private generation
	2002 Energy Policy	○ Limited role of BC Hydro in building new generation ○ Functional separation of BC Hydro, privatization of admin functions to Accenture ○ Creation of BC Transmission Corporation (reintegrated in 2010) ○ Started making large IPP power calls for new renewable generation

▶

◄ **TABLE 5.7**

Province	Policy	Details
	2007 BC Energy Plan	○ Required BC Hydro to buy private power for self-sufficiency by 2016
Nova Scotia	2004 Electricity Act	○ Mandated (private) that Nova Scotia Power allow other private generators of power (IPPs) access to the grid via an OATT
Saskatchewan	2001 filed OATT	○ Opened transmission grid to private generation
Manitoba	1997 filed OATT	○ Opened transmission grid to private generation
New Brunswick	2003 filed OATT	○ Opened transmission grid to private generation
	2004 Electricity Act	○ Expanded IPP opportunities for generation ○ Created independent system operator ○ Changed NB Power into a holding company with subsidiary structures ○ Created competitive market for wholesale, industrial, and municipal utility customers
Newfoundland and Labrador	2007 Energy Plan	○ Created NL Energy (parent company for NL Hydro)
Prince Edward Island	2005 Electric Power Act 2007 filed OATT	○ Enacted cost-of-service model of price regulation ○ Maritime Electric (Fortis) OATT approved in 2009

Source: Adapted from Blakes Lawyers 2008; Canadian Electricity Association 2010; Datamonitor 2010.

21.5 percent (see Table 5.8), all at the expense of public, rather than industrial, generation. When we compare changing ownership by source of generation, this shift is evident in nuclear, other renewables, and fossil fuels. Private actors now play a large role in nuclear generation (0 percent to 40 percent) and in fossil fuel generation (28 percent to 41 percent). Even though the percentage shift for other renewables is lower than these rates, in this emergent growth area, private and industry generation have grown from less than 1 percent to more than 15 percent in just over ten years.

TABLE 5.8

Changing public and private share of installed capacity, 2000–11

Canada		2000	2011
Total generation		111,300	129,992
	Public	90,681 (81.5%)	93,121 (71.6%)
	Private	12,777 (11.5%)	27,962 (21.5%)
	Industry	7,842 (7.1%)	8,909 (6.9%)
Renewable (hydro/ nonconventional/wind, solar, tidal)	Total	60,126 (54%)	80,042 (61.8%)
	Public	59,802 (99.5%)	67,237 (84%)
	Private	244 (0.41%)	7,654 (9.6%)
	Industry	79 (0.13%)	5,149 (6.4%)
Nuclear	Total	10,615 (9.5%)	12,665 (9.7%)
	Public	10,615 (100%)	7,655 (60%)
	Private	0	5,010 (40%)
Fossil fuels (coal, natural gas, oil)	Total	33,279 (30%)	37,283 (28.7%)
	Public	20,264 (61%)	18,227 (49%)
	Private	9,439 (28%)	15,297 (41%)
	Industry	3,757 (11%)	3,759 (10%)

Note: Numbers may not sum to totals because of rounding.
Sources: Statistics Canada 2000, 2002, 2010, 2013a.

Public Utilities: Privatizing, Unbundling, and Restructuring

Market restructuring policies are shaped by pressure from private-sector actors keen to exploit Canadian resource wealth, but also by actors mistakenly associating the private sector with innovation and efficiencies via competition (Howe and Klassen 1996; OECD 2001). In Canada, the electricity market restructuring that has occurred (most notably in Ontario and Alberta in the 1990s) was based on the argument that competitive markets and private actors are not only better able to deliver power but *necessary* for innovation, for renewables, and to avoid public debt. Policy initiatives centre on rolling back the share and scale of electricity-sector control that public utilities hold.

Electricity-sector reforms have had problematic impacts in the two Canadian provinces that have gone furthest along the standard-model road to restructuring: Ontario and Alberta. In other provinces, including Nova Scotia and British Columbia, partial reforms have led to more independent power producer (IPP) development. The Organisation for Economic Co-operation and Development economic survey of Canada from 2004 points out that "although some provinces generally consider reform of the electricity sector to be necessary, reforms have been aimed at inducing private-sector investment and protecting access to US electricity markets while avoiding full competition in generation and retail markets (e.g., establishing wholesale access and, in some cases, an open-access transmission tariff)" (OECD 2004, 1).

A Private Monopoly: Nova Scotia

Nova Scotia is unique in Canada in that a private company, Nova Scotia Power, operates as an integrated utility, running generation, transmission, and distribution in the province. Since 1992, Nova Scotia Power has been a private, wholly owned subsidiary of Emera. The company operates 97 percent of generation, 99 percent of transmission, and 95 percent of distribution in Nova Scotia (the balance is from six municipal utilities) (Blakes Lawyers 2008, 27). Privatizing the public utility did not lead to lower rates, greener sources, or more competition in the power sector in that province. In fact, changes under the province's 2004 Electricity Act were aimed at forcing Nova Scotia Power to allow other private generators of power (i.e., IPPs) access to the grid via an OATT. The act also allowed the six municipal utilities to buy power wholesale from other providers.

The majority of new renewable-power generation in Nova Scotia is procured by the utility through request for proposals. Provincial policies for renewable IPP projects did not result from a strong IPP lobby. Rather, according to one developer, "here it's more [the government's] ideological commitment combined with information from other jurisdictions" (personal interview, May 19, 2010). This unique market structure, in which a private integrated utility is the central player being forced to open to other private generation, makes restructuring in Nova Scotia somewhat different from that in British Columbia or Quebec. Although policies being enacted are similar to those in British Columbia, the incumbent utility was already both private and fossil fuel based. In December 2013, the government passed the Electricity Reform Act, which will include a major review of the electricity system in the province; if implemented as planned, these reforms will enable IPPs to retail electricity to consumers, effectively undermining the monopoly of Nova Scotia Power (Nova Scotia Department of Energy 2014).

Market Reformers: Alberta and Ontario

Arguments made by restructuring advocates that electricity competition leads to lower prices (see Chapter 4) have not been borne out by the experience in Canada so far. Restructuring was supposed to bring new generators and investment into markets, to take some of the investment pressure off public entities. The price freezes in Ontario in December 2002 had the effect of discouraging the very private investment the province wanted to create in the first place. So, in 2004, the Electricity Restructuring Act created the Ontario Power Authority (OPA). The OPA was created specifically to procure new generation via long-term power purchase agreements (Blakes Lawyers 2008, 9), since investors did not want to assume risk to build new generation. This undercuts one of the key rationales for restructuring, namely the offloading of risk (and debt) onto the private sector. Again, the record of restructuring is problematic.

Today, the argument has shifted to justifying higher costs with reference to a need for new renewables. Failing that, high prices are supposed to encourage conservation. As a result, new private generators are playing a larger role in newly opened markets in Canada, but only when the (relatively) short-term market prices are high (in Alberta) or supported by long-term government-backed contracts (as in Ontario or British Columbia). Far from a weight off the public, the costs of new generation are still borne by public actors through their taxes or higher retail prices; these costs, however, are accompanied by less direct project and planning control.

Market restructuring models promised to lead to less bureaucracy and more efficient regulation. The opposite has been true in practice. In fact, where once a single entity or regulatory body managed and oversaw the system, now many exist, each with a specialized role. For instance, one interviewee (December 2, 2010) argued that the complexity of navigating the new power system in Alberta leads to numerous problems for consumers. They are at an informational disadvantage, faced with bills that they are ill-equipped to evaluate.

In Alberta, deregulation has resulted in three regulators: the AESO (Alberta Electric System Operator – i.e., an independent system operator), the MSA (Market Surveillance Administrator), and the AUC (Alberta Utilities Commission). This structure has created more organizations where information is stored and managed, necessitating more memoranda of understanding to share information across agencies. The Utilities Consumer Advocate was also created in Alberta to help citizens understand the new market structure. It deals with the many consumer complaints about the rates and helps customers translate their bills.

In 1999, Ontario Hydro was unbundled into Ontario Power Generation (generation and retail/wholesale), with shares held by the province; Hydro One (transmission, rural distribution), a Crown corporation owned by province; the Independent Electricity System Operator (administering markets and overseeing the grid); the Ontario Electricity Financial Corporation (administering other assets and liabilities); and the Electrical Safety Authority (administering safety regulations). The 2004 Electricity Restructuring Act created the Ontario Power Authority to conduct strategic system planning and report to the Ontario legislature. In 2015 the IESO and OPA were merged. The Office of the Auditor General's 2015 report illustrates some of the reasons this merger was important, pointing to a range of failures in effective strategic planning, stakeholder consultation, and system management in the province. It makes the criticism that (among many other aspects of electricity governance in the province) "over the last decade, this power system planning process has essentially broken down, and Ontario's energy system has not had a technical plan in place for the last ten years" (OOAG 2015, 213).

In both Alberta and Ontario, the two most deregulated electricity markets, the wholesale electricity prices more than doubled (increasing from 3.1 to 8.2 cents per kWh in Ontario in 2002) (Trebilcock and Hrab 2003, 6) or tripled (from 4.3 to 13.3 cents per kWh in Alberta in 2000) (Thon 2005, 3) in the months after restructuring. This prompted the governments in both provinces to step in and control prices for retail customers – in Alberta providing a regulated rate option, and in Ontario subsidizing the rates for industry and consumers. More than ten years on, market prices in both jurisdictions remain volatile (NEB 2010a). As of July 1, 2010, the stabilized regulated rate option in Alberta was phased out; residential power rates are now based on short-term (monthly) market prices, exposing consumers to far more volatility in their power bills.

Trebilcock and Hrab point out that, even without the price increases, restructuring did not deliver on debt reduction, another cited benefit of restructuring. Indeed, the reality was quite the opposite:

> Initially, the government of the day assured electricity consumers that restructuring would lead to a reduction in Ontario Hydro's swollen debt and that the entry of private-sector companies would create competition, leading to stable, and perhaps lower, electricity prices. That did not happen ... the first year of the price freeze resulted in the issuance of approximately $730 million of taxpayer-guaranteed debt. (Trebilcock and Hrab 2003, 1–2)

What have emerged then, are volatile prices, controversies over management of grid infrastructure, increased bureaucracy, and significant opportunities for private profit. The restructured marketplace in Ontario has also led to an increase in exported power: quadrupling from 4,324 GWh and four exporting companies in 2000 to 16,656 GWh and twenty-five exporting companies in 2013 (NEB 2000, 2013).

Independent Power Producers and Piecemeal Restructuring

Alberta and Ontario have gone the furthest in Canada with their market reforms, yet other provinces are also reforming who generates, governs, and transmits power in their jurisdictions. IPP development is playing an important role in these shifts, as are piecemeal sell-offs of public utility functions. In 1996 Bruce Howe, former CEO of Atomic Energy of Canada, and Frank Klassen, former BC Hydro vice-president, argued that privatizing BC Hydro would reduce the provincial debt, improve efficiency and effectiveness, protect consumers, and eliminate political interference in utility management (Howe and Klassen 1996). Although the utility was not ultimately privatized, actors have succeeded in shifting functions to the private sector. For example, in 2003, Dublin-based Accenture took over a range of human resource and accounting functions from BC Hydro. The company argued that it would save British Columbians in the order of $250 million over ten years, which some have argued has not materialized (McMartin 2010).

In addition to the breakup of utility functions, there are mandates in British Columbia, Ontario, Quebec, and Nova Scotia that public utilities leave the development of wind, solar, biomass, and microhydro projects to the private sector (i.e., IPPs). British Columbia's experience illustrates this type of partial market restructuring via IPP development. The public Crown corporation still retains its generation assets, but the utility was functionally separated, with distinct transmission and oversight bodies created (these moves have since been rescinded). The 2001 shift to a Liberal government led to a policy mandate for BC Hydro to purchase new renewable power from private IPP sources. In 2002, IPPs selling power to BC Hydro were exempted from regulation as a public utility. Starting in 2003, several calls for power initiated a series of bids from private developers to construct, for the most part, run-of-river power plants. In the 2003 call, sixteen twenty-year contracts were awarded. In 2013, BC Hydro received nearly 20 percent of domestic electricity from IPPs, up from 10 percent in 2000 (see Table 5.9) (BC Hydro 2001, 2014). According to a 2011 review of BC Hydro undertaken by the Government of British Columbia, "in fiscal 2010, IPPs produced 16% of total domestic energy requirements; however IPP electricity costs represented 49% of the overall domestic energy cost" (Province of British Columbia 2011, 107).

TABLE 5.9
IPP purchases in British Columbia, 2000–13

	2000	2004	2008	2012	2013
Purchases from IPPs and long-term contracts ($ millions)	116	367	481	749	825
Price paid for IPP power ($/MWh)	54	59.8	61.39	69.22	74.82
IPP purchases as % of total (domestic) energy costs[1]	8	23	18 (51)	40 (73)	38 (67)
Annual payment to province (millions)[2,3]	343	73	288	230	167
% of BC Hydro domestic electricity from long-term contracts (IPPs)	10.4	12.8	12.3	18.8	18.7

1 Before 2005, BC Hydro did not separate domestic energy costs from the total in its financial statements, so only that total is reported.
2 Payments to province do not include the payments from water rentals, school taxes, grants, and capital tax to provincial and municipal governments.
3 BC Hydro pays 85% of its distributable surplus to the province. If the surplus causes the debt-to-equity ratio to exceed 80:20, it pays the maximum amount that lets it stay within this ratio.
Source: BC Hydro Annual Reports 2000–14.

These contracted purchases from private power generators, at rates high enough to guarantee them a profitable rate of return, have resulted in financial pressure on the public utility. This affects the level of annual payment that goes back to the province, as it is capped by ensuring an 80:20 debt-to-equity ratio for the utility. According to a BC Hydro 2000 report:

> The incremental cost of energy purchases is substantially higher than the embedded cost of existing hydro sources, thereby putting pressure on the gross margin. As energy purchases become a more significant portion of BC Hydro's energy supply portfolio, there will be continued downward pressure on BC Hydro's gross margin percentage. As a result, BC Hydro is facing increasing pressure to find efficiencies in other areas in order to offset the impact of increasing energy costs. (BC Hydro 2000, 47)

These efficiencies have included the outsourcing of BC Hydro staff and payroll operations, and a transfer of one-third of the utilities operations and more than fifteen hundred employees to Dublin-based global consulting giant Accenture.[6]

In Quebec, independent power producers are also playing a growing role in renewable electricity generation. The Quebec 2006 energy strategy ended the moratorium on small, privately owned hydropower stations (below 50 MW capacity). Public hydro utilities accounted for 80 percent of new installed capacity between 2006 and 2009 (the other 20 percent was industrial development) (Statistics Canada 2010, 2013b, 2014); however, between 2009 and 2013, installed hydropower in Quebec actually grew very little, with the capacities of public and private generation decreasing (by 24 MW and 185 MW respectively), while industrial generation increased by 225 MW. The majority of wind turbine development was undertaken by private companies under long-term power purchase agreements with the Crown utility, Hydro-Québec. There have been four main Hydro-Québec calls for private power: in 2001, 2008, 2010, and 2013. As a result, from 2006 to 2011, almost all installed capacity growth in wind power came from private utilities. From Mont-Joli to Gaspé, there are more than two thousand privately owned windmills stretching along five hundred kilometres, generating over 4,000 MW in total (Gagnon personal interview, May 16, 2010). Private generation from wind turbines in the province account for 88 percent of all installed wind capacity in 2011.

Green Electricity Restructuring: Public Funding, Private Ownership

The increase in private ownership across Canada is particularly apparent in the case of new renewables generation: wind power, solar, and run-of-river hydro. Electricity policy choices over the past twenty years have opened electricity generation to the private sector, particularly for new renewables generation. Consequently, the structure of wind turbine ownership across the country looks significantly different from that of the power sector as a whole. Table 5.10 illustrates the relative shares of public, private, and industry-owned wind generation. Only in Saskatchewan, New Brunswick, and Yukon is there more than 30 percent public ownership of wind power. This is because of policies, like British Columbia's, that prevent this renewable source from being developed by public utilities. British Columbia's first operational wind farm (2009), the Bear Mountain project in Dawson Creek, was being mapped for development by BC Hydro (Rison personal interview, October 14, 2009) before the public utility was directed to cede new renewables generation to the private sector. Although some provinces have initiated policies to support community ownership of new renewable power (see Chapter 6), a contradiction has emerged wherein new project development overrides community democratic control in important ways. In the case of British Columbia, the government passed Bill 30 in 2006, rescinding municipalities' right to vote on power developments in their communities. In Ontario, the Green Energy and Green Economy Act has likewise limited municipal oversight of projects by streamlining approvals (Pirnia, Nathwani, and Fuller 2011).

Wind power is not the only new renewable being developed, but it is the fastest growing. With the opening of Bear Mountain Wind Park in Dawson Creek, British Columbia, in 2009, every province and one territory now has some installed capacity. In fact, using the Canadian Wind Energy Association (CanWEA) numbers, installed capacity from 2000 to 2014 has doubled every two years. According to a CanWEA (2009) online press release, "current provincial targets and policy objectives would result in a further quadrupling of installed wind energy capacity in the next six years."[7] In 2010, this represented approximately 3 percent of installed capacity and almost 2 percent of generation in Canada (Statistics Canada 2013a). Table 5.11 shows the rate of growth of wind capacity from 1995, when there was only 25 MW, to 2014, with 8,517 MW.

Solar, tidal, and biomass sources also provide important new avenues to diversify generation, as does cogeneration from waste heat derived from

TABLE 5.10

Installed wind capacities by province or territory and ownership, 2011

	Total capacity (MW)	% of Canadian total	Public share%	Private share%	Industry share%
ON	1,741	39	24	76	0
AB	890	20	0	93	7
QC	817	18	12	88	0
NS	281	6	0	100	0
SK	171	4	93	7	0
PEI	164	4	32	68	0
NB	150	3	0	100	0
MB	104	2	0	100	0
BC	104	2	0	100	0
NL	54	1	0	100	0
YT	0.8	0	100	0	0

Source: Statistics Canada 2013a.

TABLE 5.11
Total installed wind capacity in Canada, 1995–2014

Year	MW (total installed)
1995	25
2000	139
2005	686
2010	4,022
2014	8,517

Source: CanWEA 2014.

industrial processes. None of these technologies has made a significant impact (in terms of share of GWh) in the generation of electricity in Canada so far, but all may – and should – play a key role in coming decades. Whether and where they do depends on energy and environmental policies that are shaped by an important range of actors, which include incumbent industries as well as the environmental and community groups discussed in Chapter 8 and the international forces in Chapter 4.

Established Energy Lobbies
The fossil fuel lobby has a long and well-documented history of shaping Canadian public policies at federal and provincial levels (Doern and Toner 1985; MacDonald 2007; Newell 2000). Unlike environmental NGOs (ENGOs), businesses in upstream (mining, drilling) and downstream (refining, processing, and distributing) areas of fossil fuels have frequent and direct access to political and regulatory bodies as part of their regular business activities (Bratt 2006; Murphy and Murphy 2012). These relationships with provincial power authorities, planning agencies, environmental resource departments, and elected leaders establish an important conduit for the preferences of incumbent fuel businesses to shape policies. This impact is supplemented by public lobbying efforts at times when private negotiations prove insufficient to clear the way to continued business development (Curry and McCarthy 2011; MacDonald 2007; McLeod-Kilmurray and Smith 2010; Sinoski 2012).

Established industry actors in areas of key strategic economic significance for the country enjoy a privileged – and difficult to dislodge – role in policy making. Unsurprisingly, when their core business area is threatened by new policy actions, these actors leverage both private and public mechanisms to delay and, if possible, reverse negative outcomes. Given the diversity of ownership and fuel sources across the country, the specific configuration of actors and sources looks rather different in Alberta or Saskatchewan than it does in Ontario or Quebec. In Alberta, the dominance of fossil fuels and private industry leads to a particularly powerful and closely wedded relationship between the government and private drilling, refining, and distributing businesses (Bell and Weis 2009; Hoberg 2011). In Ontario, the key energy actors consist of a mix of public and private nuclear, coal, oil, gas, and hydro entities.

In the electricity sector, the nuclear and coal lobbies have played an important role in opposing aspects of new renewable-energy policies, and in limiting the penetration of new actors (Etcheverry 2013). This is most clear in the case of Ontario's nuclear industry, which has played a central role in limiting grid access available for co-operatives. The nuclear industry is economically important, especially for high-skilled, high-paying employment in Ontario: thirty thousand people are directly employed at nuclear plants, and more than twice that indirectly employed in mining, transport, and other related areas. The industry also pays $1.5 billion in federal and provincial taxes (Bratt 2012). It has been more successful than the coal industry in maintaining its place in the electricity mix: whereas coal's share in the power sector has moved from 25 percent in 2003 to zero in 2014, nuclear still provides half of the province's electricity. This is partly because of the heavy concentration of nuclear facilities and industries in Ontario, creating a powerful and very active lobby group.

The stakes of nuclear lobbying efforts are high today, given the opportunities that climate challenges present for enhanced public investment but also the drawbacks that accompany public relations disasters like 2011's Fukushima Daiichi meltdown. Bratt (2012, 21) notes that "if the provinces move away from nuclear power and either maintain coal and natural gas power plants or expand their use of renewables, they will kill the nuclear industry in Canada. And if they choose a reactor design other than CANDU, that choice will lead to a fundamental restructuring of the Canadian nuclear industry." The position of the Organization of Canadian Nuclear Industries (OCI) is that "our provincial and federal governments are under constant pressure from antinuclear groups. The OCI holds regular discussions with government officials at various levels to promote our industry and its many benefits to Canadians" (Bratt 2012, 23).

The nuclear industry is also characterized by a closely connected and well-organized network of actors. These actors include the Canadian Nuclear Energy Association, which acts as the voice for the industry with Canadian government. It has more than a hundred members, including Atomic Energy of Canada limited (AECL), suppliers, reactor operators, provincial power utilities, medical isotope producers, universities, and scientific associations. AECL is a Crown corporation, with five thousand employees headquartered in Ontario. As of 2011, global energy giant SNC-Lavalin owns the reactor division of AECL. Ontario Power Generation is a provincial utility that owns and operates reactors at Pickering and Darlington. It also owns but leases out operation of the Bruce nuclear stations to the Bruce Power Limited Partnership. Bruce Power's major owners are Cameco (one of the world's largest uranium companies), TransCanada Corporation (a Calgary-based pipeline and generation company), and BPC Generation Infrastructure Trust (a trust of Ontario municipal employees). The OCI promotes export markets for the nuclear industry, and the Canadian Nuclear Workers Council represents the interests of nuclear-sector employees.

One of the other reasons the nuclear lobby has been more successful in Ontario than the fossil fuel lobby is that the Nanticoke coal-fired generating station was one of Canada's top ten GHG sources. It played a very visible, high-profile role in contributing to air pollution as well as to greenhouse gases. Nuclear power, by contrast, does not directly emit greenhouse gases in the generation of electricity. Of course, many environmental advocates are quick to point out that other parts of the process – mining of uranium, enrichment, transport, and construction – do. They also point out that the persistent issue of nuclear waste storage has yet to be solved and that new nuclear energy projects are exceptionally costly compared with other available options (Godoy 2011; Martin 2012). It is, however, a more complex environmental challenge than air pollution, as the issue is spread over long time horizons and, for many, lacks the immediacy of other environmental issues (Durant and Johnson 2009; Johnson 2004). Certainly, disasters like the Fukushima Daiichi meltdown can bring the risks of nuclear into sharp relief, but actors remain split on the ultimate role the source should play in any green energy future.

What emerges from this brief outline is that the relationship between incumbent industries in renewable transitions is extremely important. In Denmark, a powerful social movement pushed widespread restructuring and was able to counter more established forces. Germany, in the wake of the Fukushima disaster, committed to decommissioning nuclear plants by 2022. In Ontario, the story has been different. The size and scale of the nuclear generating plants have meant that the province has restricted the ability for new renewable-power generation to be fed into the grid in areas near these large thermal facilities. A clear tension exists between incumbent nuclear and coal generators defending centralized power generation and priority grid access, and new renewable-energy advocates pushing to undermine these power sources and shift to wind, solar, and hydro generation.

Ontario's Green Energy Act

In Ontario, recent changes to the power sector through the Green Energy and Green Economy Act have been both successful and contentious. They clearly illustrate the issues with using the private sector as a vehicle for new renewables development, as well as the incumbent actor's power. These issues are significant because other provinces, including Nova Scotia and New Brunswick, are looking to Ontario as a model for greening their power sectors (Nova Scotia Power personal interview, May 19, 2010). This growth of new renewables was necessary, in part, because of the province's bold commitment in 2007 to phase out its 18 percent installed capacity share of coal-fired generation by 2015 (the goal was achieved a year ahead of schedule in 2014). The electricity restructuring in the late 1990s and the opening of the competitive electricity market in 2002 had clearly failed to lead to significant new renewable generation, as market upheaval, incumbent actors, and profitability considerations dominated the landscape. As a result, renewable incentive programs became necessary to encourage private actors to develop new and often more expensive sources of power. Since their inception, these programs have undergone continuous renegotiation and change in response to a range of debates over system capacity, appropriate pricing, effectiveness, and most recently, compliance with international trade rules (Stokes 2013a).

Ontario has been experimenting with renewable incentives since 2006, when it initiated its Renewable Energy Standard Offer Program (RESOP). This program provided generators with long-term fixed-price power purchase agreements with the public Ontario Power Authority. To spur new (to Ontario) generation technologies, the system was set up to provide a higher guaranteed tariff (rate) for solar projects than for wind, biomass, or other renewable sources. This price differentiation by source is also known as a feed-in tariff (FIT) program. The RESOP was the first FIT program of its kind in North America. FITs are statutory arrangements that set prices for renewable sources. The price set for FITs is based on a determination of "reasonable profitability" by policy actors and is generally described as the price of generation plus a reasonable return. If a project meets the criteria specified by the power authority, it is granted a contract. Unfortunately, the RESOP did not result in the targeted level of new renewables, either from communities or from the private sector (Holburn 2012).

Although the RESOP provided guaranteed and differentiated rates, it was not considered a full FIT, since there were only two source rates and it failed to include differentiation by project size. Critics argued that tariff rates needed to be designed to provide a return based on the differential costs and benefits of both source and scale. In 2009, Ontario's government passed the Green Energy and Green Economy Act, which fully expanded the range of tariffs to include reasonable returns based on many diverse generation technologies and project sizes. This included a shift from two rates to nineteen, with more than four separate rates for different sizes of rooftop solar projects and significant increases in the rates for solar generation (see Appendix 6). The new act also included local content requirements for solar and wind projects on the basis that these would help stimulate a green-energy economy in the province (OPA 2009, 2010, 2013a). It also included a 1-cent-per-kilowatt-hour adder (extra payment) for community-owned projects, and 1.5 cents for Aboriginal power.

Ontario was not alone it its selection of this policy tool. FITs are increasingly being applied around the world. They are the key policy choice in place in some US states (California, Hawaii, Vermont), as well as in Germany, Denmark, Spain, and more than thirty other countries around the world (Barclay 2009; Gipe 2007; Lipp 2008). The relative successes in the development of wind power in these jurisdictions has led others to look to the FIT model as a best practice for new renewables, particularly community renewables (see Chapters 6 and 7). FITs are also seen as more effective at actually getting new projects built, and are, based on experiences in Germany and Denmark, more favourable than other market-based procurement mechanisms to small (co-operative and community) IPPs. By many measures, the program in Ontario has been a success: coal-fired plants have closed a year ahead of schedule, the Ontario's target of 10 percent to 15 percent new renewable generation by 2018 will be reached five years earlier than planned, and billions of dollars of private investment have flowed into renewable electricity generation (Ontario 2012, 2013a, 2013b).

Regular bi-annual program reviews and annual price reviews provide policy makers with an opportunity to adjust FIT rules in response to stakeholder consultations and changing market conditions. In 2012, FIT 2.0 – the first review – rules included reduced tariff rates of up to 32 percent for solar and wind projects and two more size categories. It also established a priority points system for Aboriginal and community participation projects and a 10 percent set-aside for these projects. See Table 5.12 for an overview of the changing FIT rates from 2006–14. Other significant changes have taken place in recent

TABLE 5.12
Sample Ontario RESOP and FIT rates, 2006–14[1]

	2006	2009	2012	2014
Wind (250 kW)	11	13.5	11.5	11.5
Solar photovoltaics (250 kW)	42	71.3	53.9	32.9
Solar groundmount (250 kW)	42	n/a	n/a	28.8
Hydro (1 MW)	11	13.1	13.1	14.8
Aboriginal 50% participation	n/a	1.5	1.5	1.5
Community 50% participation	n/a	1	1	1

1 This is only a small sample of the range of rates, sources, and project sizes from 2006 to 2014. For a full list, see Appendix 6.
Sources: OPA 2005, 2009, 2012, 2013a, 2014.

years. (For example, FIT 3.0 in 2014 effectively cancelled the FIT program for projects over 500 kW and implemented a simplified FIT and microFIT for small projects [Gipe 2013]. In January 2015 the Ontario Power Authority merged with the IESO and launched a new separate process for large renewable procurement, with the first results announced in March 2016.)

Ontario's policies have also been contentious. First, while IPP supporters justify these moves on the basis that private investments help shelter governments and ratepayers from financial risk, they neglect to point out that system coordination, infrastructural upgrading, and profit-based rates are costs borne by ratepayers and the public sector (Calvert 2007; OAGG 2015). The Ontario FIT in particular has been criticized for being excessively expensive and having a negative impact on social welfare in the province (Pirnia et al. 2011). IPP agreements with integrated utilities are guaranteed purchase contracts at high prices (10 to 80 cents per kWh) paid by households, many times the cost of conventional (older) power generation. Although the FIT program in Ontario has increasingly prioritized community and First Nations projects, the overwhelming majority of projects and corresponding financial returns still accrue to large energy companies. This has drawn criticism from both sides of the political spectrum (Gipe 2013; Stokes 2013b).

Second, FIT rules continue to emphasize the importance – through allocation of priority points as well as contract requirements – of community and municipal planning consultations. This is a result of a backlash against the first FIT rules, which reduced the role of municipalities in project approval, in an attempt to facilitate new project development. This, together with the opposition to large wind developments that have taken place in many jurisdictions around the globe, prompted the energy ministry to take issues of local opposition more seriously. One community actor expressed how this change has provided vindication for the community sector's efforts over the past two decades:

> It has taken years and years ... as a sector we have been treated like children much of the time. We've got a foothold now, and at least the current government recognizes our value and that they've made mistakes in not paying closer attention to the impact of getting thousands of private citizens involved in an ownership capacity. This is certainly the case looking in hindsight at some of the planning objections and backlash in Ontario with wind. (Personal communication, July 27, 2009)

One example of these changes is that, in FIT 2.0, projects with evidence of municipal support and those with local ownership gain priority access to the contract queue. FIT 3.0 changes included a new category of public or municipal participation alongside the adders for Aboriginal and community participation, and increases the minimum qualifying participation rate from 10 percent to 15 percent.

Finally, the Green Energy and Green Economy Act included local (Ontario) content requirements for wind- and solar-project components in order to qualify for the FIT. In 2010, however, Japan and the European Union launched a complaint against Canada (Ontario) at the World Trade Organization, alleging that "under these measures, technologically advanced and highly competitive and sophisticated solar panels or other renewable energy generation equipment produced in Japan are discriminated against ... simply because of their origin" (Agence France-Presse 2011). In May 2013, the World Trade Organization ruled against Canada, finding that the local content requirements violate the principle of national treatment (Howlett, Marotte, and Blackwell 2012). In particular, the ruling found that because electricity was not procured by the government for governmental use, but rather for commercial resale, the program was not exempted from the national treatment clause (WTO 2013a, 2013b). In June 2013, Ontario's energy minister, Bob Chiarelli, announced the province's intention to bring the program into compliance by eliminating the local content requirements, as well as the end of the FIT program for projects larger than 500 kW. These large projects are now subject to a competitive bidding scheme (REN21 2014).

However, for community energy advocate Paul Gipe (2013), the most recent changes announced "relegate feed-in tariffs to a small-project ghetto of projects less than 500 kW in capacity. [Since] no commercial wind energy can be developed for less than 500 kW when each wind turbine is 2 MW or more ... community wind is dead." In its 2014 annual report, the Federation of Community Power Co-operatives echoed a disappointment with the planning rules, arguing that "unfortunately, the LRP [Large Renewable Procurement] rules, which were released in March [2015], did not reflect any of the comments we had submitted regarding the importance of offering communities meaningful participation and it is likely that there will be no co-op participation in any LRP projects. It is unfortunate for the province as the resistance to wind is likely to continue" (FCPC 2015, 1).

The Ontario case is now a cautionary tale of policy makers seeking to increase the local multiplier of benefits from renewable energy projects via "green growth" policies (Howlett, Marotte, and Blackwell 2012; Lord 2011; Wilke 2011). Similar trade challenges between the United States and India were filed in 2013, and numerous other major economies have similar local content requirements in their own energy policies under review (Kuntze and Moerenhout 2013). These recent developments are concerning given that local economic benefits will be necessary to form and sustain the political support for ambitious renewable energy transitions.

Greenwashing Power and Profit

Environmental forces for change in the electricity sector have political economy implications but biophysical causes. Neoliberal forces, on the other hand, have political economy causes but biophysical implications. If we are to tackle the environmental implications of power, we need to address political economy issues of ownership and distribution. The scale of restructuring needed to shift the Canadian economy off an environmentally and socially self-destructive path requires systemic and radical change – requires taking on powerful actors and industries, reshaping prices and consumption preferences (Daly 1996; Faber 2008; MacArthur 2014). Market-based environmentalism is simply not up to the task; the degree of environmental reform needed to deal adequately with climate change requires strong coordinated intervention across industrial sectors by the federal and provincial governments, as well as significant infrastructural spending on grids and generation (Jaccard and Simpson 2007; Weis 2010; Weis et al. 2009). Without this coordinated investment and intervention, policy targets are far more likely to lead to *greenwashing*, wherein companies and governments spin policies as environmentally friendly to appease public opinion while continuing to degrade the environment.

In the case of electricity, claims about the virtues and consumer benefits of privatization and deregulation are overstated. The real work of infrastructural upgrading, of providing incentives for new renewables and public education, is borne by the state. For high-GHG-intensity provinces, improving the environmental record of the power system means taking on the issue of source of generation. For the provinces with higher hydro capacity, the challenge is to work much harder on demand management and, where necessary, diversify to include new renewable technologies. Canada, as an electricity system dominated by relatively low GHG emissions hydroelectric power, is in some provinces doing quite well in comparison with other states around the world. In the United States, for instance, two-thirds of all power generation in 2014 came from fossil fuels (39% coal and 27% natural gas). This is not to say that diversification through the introduction of wind, solar, and biomass is not useful or important, merely that for provinces with very low GHG intensity, the benefits of demand management and reducing power consumption are a bigger part of the puzzle than is shifting electric power away from existing hydro facilities to other (new) sources of generation.

As it stands today, generation from new renewables like tidal, solar, and wind account for a very small share of total generation in Canada (less than 2 percent in 2014). Almost all new growth in this sector is private generation. The pairing of green power with private power raises serious concerns. If the goal of the utility is not low stable rates but profit, the utility has little incentive to reduce consumer demand or to invest over the long term. Indeed, electricity rates across the country have been rising steadily over the past ten years. Some of these costs are necessary, such as upgrading aging infrastructure. However, some are intimately tied to the push for private accumulation and expanded continental grids. This has led to double-digit electricity rate increases in Nova Scotia and New Brunswick (NEB 2010a), as provincial actors provide incentives for new generation sources and to tackle aging infrastructure.

In Ontario, the liberal government rates have risen 50 percent since 2000, in significant part to address the costs of nuclear and transmission systems as well as the phase-out of coal in 2014. New higher rates are not just going toward upgrading, improving and bringing greener generation sources into the system, but also toward enriching investors at home and abroad, and the process of system restructuring in Ontario has recently been criticized in the Ontario Office of the Auditor General's Annual Report (OOAG 2015). Alternative modes of development (in this case, public renewables) oriented more strongly toward public needs are crucial as well given the increased costs associated with sectoral reforms.

Table 5.13 shows the changes in average residential utility rates between 2000 and 2015. In British Columbia, rates are set to rise by a further 29 percent by 2019 (British Columbia 2013). This is because of the increasing costs of IPP generation together with infrastructural upgrades like smart metres and subsidizing transmission lines to support mine development (in the case of the Northwest Transmission Line project). These price increases in British Columbia are out of step with fluctuations in the economy more broadly and are problematic given the centrality of the power sector to the economy (Statistics Canada 2010, 20). Fuel poverty for low-income Canadians is a very real risk in coming years.

A confluence of pressures is thus leading to rate increases across Canada. These include the need to upgrade generation to greener sources and to build new transmission for system reliability and export. They also include pressure to generate electricity for profit and exchange rather than for local consumption. Each of these pressures raises problems of democratic control and scale. The expanding continental market undeniably generates revenue for some communities, provinces, and corporations. What is sacrificed

TABLE 5.13
Residential electricity rates in Canadian cities (¢/kWh)

City	2000	2015	% change
Vancouver	6.41	10.29	61
Ottawa	10.09	14.86	47
Halifax	11.21	16.03	43
Regina	10.43	14.37	38
Charlottetown	12.15	15.62	29
Winnipeg	6.30	8.11	29
Toronto	11.14	14.31	28
Moncton	10.14	12.30	21
St. John's	9.88	11.55	17
Edmonton	10.22	11.55	13
Montreal	6.60	7.19	9

Source: Hydro-Québec 2001, 2015.

is local and public control over how and where this energy is produced, not to mention how much it will cost. In this vein, Griffin Cohen argues that

> powerful trade agreements that support an export-centered energy strategy can compel markets to open in ways that will jeopardize the stability of both supply and pricing that Canadians take for granted ... The major risk for Canadians in a deregulated market is that the new private producers, who will have access to the transmission grid, will focus on exporting to the more lucrative market in the United States. Since public utilities would no longer plan for future supply, but rely on the private sector's investments, and since prices would no longer be regulated to reflect the cost of production, Canadians would be forced to compete with customers in the United States for access to their own domestically generated electricity. (Cohen 2004, 6)

International export orientation and continentalism, together with the well-organized interests of the nuclear sector, impact long-term sustainability in the electricity sector. According to Cohen (2004, 4), "this [orientation] has produced unfortunate results, such as Ontario developing nuclear power rather than importing significant amounts of hydroelectric power from Quebec, and Alberta relying on coal rather than importing much hydroelectric power from British Columbia or Manitoba." These trade-offs and trends are important for the prospects of co-operative and community-based electricity in Canada. The power sector is scaling *up*, not down, and orienting grids to distribute power less efficiently across greater distances.

The way in which the shift to renewables is taking place across Canada ultimately undermines our power in both senses of the word; that is, it undermines the average Canadian's access to electric power, and it undermines Canadians' power to properly manage the transition to a greener future in a meaningfully democratic way. Energy is not just a commodity for sale. Access to electricity and control over its sources (e.g., for environmental reasons) is a matter of citizenship (Doern and Gattinger 2003; Hampton 2003). Neoliberal power-sector reforms are eliminating a critical tool for provinces and territories to protect the environment, create new technologies, help manage demand, and provide jobs and low-cost access to power for poor families (Byrne, Toly, and Glover 2006; Hampton 2003). In other words, these reforms are undermining the security *and* sustainability of the power sector in Canada.

Summary

Neoliberal restructuring of the power sector was not driven primarily by technical or environmental challenges; it is not inevitable, but contingent on currently popular rhetoric about the virtues of private markets. The direction of power-sector reforms taking place in Canada today is contradictory and problematic. These forces of change lead to contradictions between public expectations and desires for greener and more secure power and the neoliberal practice of electricity production and regulation that favours privatization and continentalization. The implication of the emerging continental market is that Canadian communities and localities in exporting provinces may soon be paying the costs of supporting private-sector expansion. Moreover, the profits from sale and production are not necessarily circulated back into the communities bearing the environmental costs of generation and transmission (for generations). Transmission systems come with environmental and monetary costs borne by communities in spatial proximity, whereas the benefits accrue to power traders and energy investors. If sustainability and the mitigation of climate change are a serious goal – as they should be – these developments need to reverse.

In contradiction and change there is opportunity. New generation technologies (small-scale, hydro, wind, and solar) are environmentally desirable and economically possible. Electricity reforms open up space for private (and more local) actors like co-operatives, and decentralization of generation. The open access to transmission grids and contracted support for new renewables facilitate, in theory, a range of new actors, including community actors. In Ontario, co-operative and community groups have organized to develop projects, thought impossible only fifteen years ago (see Chapters 6 and 7). What is problematic, however, is that although often cited in rationales for restructuring, community projects can be a rhetorical Trojan horse or "communitywash" that sanitizes a shift of electricity ownership to large (typically) private actors (Gagnon personal interview, May 16, 2010; Catney, MacGregor, et al. 2013). In fact, despite enthusiasm and hard work toward building community energy systems, empirical evidence from across Canada demonstrates that they are unlikely to form a major part of new power developments given the power of incumbent energy industries.

6

Electricity Co-operatives
The Power of Public Policy

Electricity co-operatives are not new in Canada. In the 1940s and 1950s, provincial governments in Alberta and Quebec used electricity co-ops as alternative service providers. This history provides important lessons for the future of new co-operatives in the sector. One challenge in studying the role and potential for power-sector co-operatives is that very little has been written about their earlier development. In what follows below, Canadian electricity co-operatives over the past seventy-five years are profiled, as are specific provincial policy initiatives that have guided co-operative development. Provincial public policy has been and continues to be a central driver of electricity co-operative development; in the past this was a result of provincial governments being unwilling, for ideological reasons, to spend public funds to electrify rural areas. That is, provincial politics together with public-policy support for co-ops and rural mobilization explain their development. New electricity co-operatives are emerging as a result of neoliberal policy shifts that encourage private participation in the power sector, along with increased interest in environmental issues.

Since 1990, at least 165 electricity co-operatives have incorporated in Canada, and 715 since 1940. This number is likely significantly understated due to the rapid incorporation of these co-ops in Ontario in the past two years. Electricity co-operatives can generate electricity, distribute it, or be formed by groups of consumers to secure better rates or particular services. Important differences exist in the forms, locations, and motivations of contemporary Canadian power co-operatives. Data on incorporation rates and co-operative profiles from two periods is compared below: 1940 to 1990 (development) and 1990 to 2013 (restructuring). In the development phase, co-operatives focused their efforts on building access to electricity and services for underserved rural communities. This, for the most part, meant co-op distribution (power lines). Nearly all early co-operatives developing during this time incorporated before 1970. This was largely because of the policy supports provided. It can also be attributed to the postwar policy context wherein infrastructural development in the electricity sector was still developing, especially in rural areas. During the 1990–2013 period, co-operative development changed in form because the challenge was based on rising prices and community concerns with environmentally damaging generation sources, rather than on providing rural access.

This historical data provides invaluable comparative background for new co-operatives. Electricity co-operatives today are less numerous, are just as likely to be urban as rural, and are spread more evenly across the country's regions. Unlike the distribution co-operatives of the mid-twentieth century, power co-operatives today are in a newly profitable niche, and this has consequences for the kinds of contributions that co-operatives can make going forward (see Chapters 7 and 8). Community power-policy initiatives have arisen in the past seven years to facilitate co-operative electricity generation. These policies use a range of policy tools to facilitate community power development. A marked lack of success – or sometimes even expectation of success – is present for co-operative project developments in all provinces but Ontario. Even there, the purchase in the larger power sector remains small. Incorporation has not led to developed projects in most cases. The range of challenges contemporary electricity co-operatives face is addressed throughout the rest of this book.

Types and Concentration of Electricity Co-operatives

There are four main types of co-operative activity in the electricity sector. These are, in order of prevalence:

1 Electricity *distribution* co-operatives, where members pool assets to build (or buy) sections of the distribution grid.
2 *Generation* co-operatives that own generation assets and sell power to public utilities or private retailers.[1]

3 *Consumer retail* co-operatives that purchase bulk electricity and resell to members at a lower cost than regular market rates – possible in deregulated retail markets like Ontario and Alberta. Consumer co-operatives can also provide services to members to conduct energy audits and install residential electricity-efficiency equipment.

4 *Networking* co-operatives, which serve as community associations to provide members with an avenue to promote renewable electricity and policy change.[2]

Within each of these four main co-operative activities are various co-operative ownership structures, including worker co-operatives, examined further in Chapter 7. Table 6.1 provides a numerical summary of co-operative development in the power sector in Canada from 1940 to 2013.[3]

Note the significant difference in number of new incorporated co-operatives between these two periods: 549 in period 1 (development) and 166 in period 2 (restructuring). This difference is not a function of the extended length of period 1, as only six incorporated in the two decades between 1970 and 1990. More rural utility co-operatives were started between 1947 and 1957 in the province of Alberta than have been incorporated in any area of the electricity sector in all provinces put together since. It may, however, be too early to make these comparisons, as a new cycle of co-operative development has just begun, with eighty-eight new incorporated co-operatives in Ontario alone since 2010.[4] The provincial initiatives in 2009, 2010, and 2011 driving this are outlined later in this chapter. It may, therefore, be useful in the future to revisit and compare the data above with the 1990–2040 period, to assess both in fifty-year blocks (1940–90 and 1990–2040). That said, what is presented in this chapter is the most comprehensive account to date comparing electricity co-operative development across the country.

TABLE 6.1
Electricity co-operatives by type incorporated, 1940–2013

	1940–90	1990–2013	Total (type)
Distribution	542	19	561
Generation	3	129	132
Consumer	3	7	10
Networking	1	11	12
Total (time period)	549	166	715

Sources: CCA 2011a; Co-operatives Secretariat Survey 2010; Doiron 2008; FSCO 2014; project and association websites.

TABLE 6.2

Electricity co-operatives by province and period incorporated, 1940–2013

Type (total #)	BC	AB	SK	MB	ON	QC	NB	NS	PEI	Canada
1940–90	1	381	1	-	3	163	-	-	-	549
1990–2013	2	20	2	3	105	21	8	4	1	166

Sources: CCA 2011a; Co-operatives Secretariat Survey 2010; Doiron 2008; FISCO 2014.

Electricity co-operatives are geographically concentrated within each period. Between 1940 and 1990, development was concentrated in Alberta (70 percent) and Quebec (29.5 percent). Between 1990 and 2013, Alberta's share of new incorporated co-ops dropped to 12 percent, and provinces that didn't have any – Manitoba, New Brunswick, Nova Scotia, and Prince Edward Island – started developing them. Two forces have played a role in shifting concentration of new power co-ops from Alberta to Ontario, among other provinces: supportive policies for community power generation and public frustration, particularly among environmentalists in more coal-reliant provinces (Nova Scotia, Ontario, New Brunswick, Alberta) with progress toward developing renewables.[5] Table 6.2 illustrates this geographic shift.

Early Development: Distribution Co-operatives, 1940–90

Ninety-nine percent of electricity co-operatives in Canada on record prior to 1990 distributed power to rural residents. These co-operatives emerged following the Depression, during the interwar period. They played an important role in rural electrification. The failure of private-sector rural electrification and consequent rural mobilization are what drove them. These factors were ultimately instrumental to co-op incorporation. Provincial choices to support co-operatives reflected both policy spillover from the United States and the ideological leanings of provincial governments at the time. In the United States, President Roosevelt's New Deal policies – the establishment of the Rural Electrification Administration in 1935, in particular – promoted co-operative development for rural utilities through grant funding and legislation granting franchise areas.[6] Private companies were opposed to these programs, arguing that public competition was unfair, as were subsidies and supports for co-operative rural electrification. Without these programs and facing an industry that would provide access only for a profit, rural people were left without electricity.

In Canada between 1940 and 1990, co-operatives incorporated in the electricity sector to build distribution grids in Alberta (381) and Quebec (162).[7] Not all of these succeeded due to a combination of financial and technological issues (Doiron 2008; Dolphin and Dolphin 1993). In other provinces, by contrast, a combination of public, private, and municipal utilities developed the electricity distribution systems and, in the 1960s, provincial public utilities bought out private and smaller local and municipal utilities (Dupré, Patry, and Joly 1996; Netherton 2007). By 1967, provinces ranged widely in degree of public control of electricity assets, from 99 percent in Manitoba to 4 percent in Prince Edward Island (Dolphin and Dolphin 1993, 6).

Policy Supports in Quebec and Alberta

Quebec and Alberta diverged from other provinces because conservative governments were in power at the time when farmers and rural communities mobilized to gain electricity access. Both provincial governments looked to the US experience of rural electrification in deciding how to bring power to rural areas and avoid the public-sector ownership that had emerged in other provinces (notably Ontario and British Columbia). The ideological opposition of both Maurice Duplessis's Union Nationale (1944–59) and Ernest Manning's Social Credit (1943–67) to public ownership of power systems, combined with popular – albeit fragmented – pressure for action, led to the support for the co-operative "third way" (Doiron 2008; Dolphin and Dolphin 1993). This approach allowed for some service provision with minimal outlay of public funds. A key difference lay in the fact that the ideological winds changed in Quebec, and did not in Alberta. As a result, Quebec bought out the co-operatives and brought them into Hydro-Québec during Jean Lesage's Liberal government and the Quiet Revolution in the 1960s.

Rural electrification co-operatives were most active in Quebec between 1945 and 1963 (less than twenty years). One hundred and sixty-two rural electricity co-operatives incorporated (Doiron 2008, 135),[8] served more than eighty thousand people, and covered more than twenty-four thousand square kilometres. The earliest electricity distribution co-operative in Canada was established in Compton, Quebec, in 1939 (Saint-Pierre 1997). It borrowed money from a local credit union to build a power line to connect to Southern Canada Power's network and electrify a small agricultural area. It lasted until 1961 (Doiron 2008). Just a few years later, in 1944, the Co-opérative régionale d'électricité de St-Jean-Baptiste-de-Rouville incorporated in Quebec; it is the only electricity distribution co-op on record in

Quebec that still operates today. The co-operative developed as part of the 1945 Rural Electrification Act in Quebec established by the Duplessis government (Doiron 2008; Dupré, Patry, and Joly 1996).

According to Hydro-Québec, the Rural Electrification Act contributed to the modernization of Quebec farming in the postwar period (Hydro-Québec 2011). It created a $12 million fund and the Office de l'électrification rurale du Québec (OER) to help rural electricity co-operatives develop via the provision of technical, legal, and financial assistance. Two hundred and sixteen municipalities requested rural electrification funds to start rural electricity co-ops, the bulk of which were formed in 1945. Between 1945 and 1948, $2.1 million (in 2008 dollars) in funds was advanced to co-operatives (Doiron 2008, 184). Among other things, the act mandated that distribution co-ops build and manage power lines, with the supervision of the OER. In 1963–64, all but one of the forty-five remaining rural electricity co-operatives in Quebec were bought by Hydro-Québec and integrated into the public distribution system. The Co-opérative régionale d'électricité de St-Jean-Baptiste-de-Rouville was the sole holdout and now distributes electricity to sixteen municipalities in the Montérégie region.

In Alberta, local need and provincial policy spurred distribution co-operatives. In 1941, rural Albertans had only a 5 percent electricity connection rate, as opposed to 33 percent in both British Columbia and Ontario (Meyer 2009). The Alberta government's reluctance to build a public power system, despite the push from rural farmers, played a key role in co-operative development. According to Ernest Manning, Alberta's premier during the 1940s, public power was socialism (Dolphin and Dolphin 1993, 26). Living in one of the last provinces to enact rural electrification policy, rural Albertans were largely left to their own devices, which is to say, they largely lived without power access. Private gas and electrical utilities were unwilling to extend access to underserved areas, as profit margins were far lower per kilometre of line or pipe, leaving these populations to fend for themselves. In 1948, a provincial plebiscite was held on whether Alberta should take over the power sector or the private companies should continue with business as usual. The private option won, but by fewer than 151 votes (Dolphin and Dolphin 1993). That left farmers with only the Co-operative Marketing Associations Guarantee Act of 1946, allowing for government guarantees of up to half the construction costs; farmers needed to secure the balance of funds privately.

Following the establishment of Alberta's rural electrification program (in 1947), over 90 percent of Alberta farms were connected to power within a decade (Dolphin and Dolphin 1993, vi). These co-operatives involved buy-in shares of $100 for rural farmers, with the balance of funds to build the system to be borrowed from banks, the loan subject to a provincial guarantee. Unlike Quebec, Alberta's rural communities bore most of the cost themselves. According to Pat Bourne, CEO of the Central Alberta Rural Electrification Association (CAREA) (personal interview, December 1, 2009), in devising the co-op policy, "the large investor-owned utilities were involved [in the system development], which was unfortunate because we didn't get franchise areas because of that. We got what's termed a 'service area.' We share the service area with an investor-owned utility." This lack of franchise area has played a key role in the constant decline in Alberta's power distribution co-operatives in subsequent years.

The relative success of co-operatives in meeting rural electrification and community development needs, as measured by longevity and penetration of co-operatives in the energy sector in Alberta, provides a historical basis for future co-op development. It also brings with it warnings. These co-operatives scaled up specifically because public policies facilitated – and in some senses forced – their growth. Certainly, local mobilization, time, and entrepreneurial effort played a role (Dolphin and Dolphin 1993; Yadoo and Cruickshank 2010). But these efforts were financially supported by provincial policies that used public funds to develop private power. One lesson that may be taken from this history is that both political mobilization and supportive public policy are essential for broad co-operative development. In the electricity sector, this may be especially important given the significant competition, the infrastructural investments required, and the importance of balancing the playing field with investor-owned utilities.

A spillover effect of earlier co-operative development also led to the development of other energy-sector co-operatives. First in the United States and then in Quebec and Alberta, co-operative supportive policy provided a solution for service provision in a (then) unprofitable part of the power sector. In addition to this international policy learning, the success in one co-operative network, electricity, had spillover effects to others in natural gas. Rural gas co-operatives began constructing distribution networks in the 1950s as a result of the electricity co-operatives' inroads. In 1970, there were twenty-five small rural gas co-operatives in Alberta. This was before the 1973 Rural Gas Act, which allowed gas co-ops to scale up by providing grant assistance for construction of rural gas-distribution systems and setting up Gas Alberta to procure gas and franchise areas to secure customers. The number of new gas co-ops grew rapidly through the 1970s, and by 1979, fifty-nine were incorporated in Alberta. These co-operatives learned from the experience of earlier electricity co-ops, so they pushed and were granted exclusive franchise areas in the province. As a result, they are much better resourced today than the electricity co-operatives (personal interview, November 27 and December 1, 2009; Dolphin and Dolphin 1993; Orr 1989).

Restructuring: Generation and Renewable Energy Co-operatives, 1990–2013

Electricity co-operatives incorporated since 1990 differ in both function and geographic distribution from those of the earlier era. Over the past twenty-five years, electricity co-operatives have shifted mandates and are increasingly forming in order to provide other services: generation of renewables, networking, and retail of renewable energy products such as solar panels. In some cases, they are also forming consumer electricity-purchase co-ops. These new co-operatives in the power sector are, as of 2015, less geographically concentrated and do not have guaranteed rights to either government funds or designated franchise or service areas. At the same time as new types of electricity co-operatives are emerging, the remaining rural distribution co-operatives in Alberta are under pressure to sell to investor-owned utilities (IOUs). These changes in the form and number of co-operatives are driven by a restructuring of the power sector in provinces across Canada, as outlined in Chapter 5, including opening up the grid to new private-generation retail opportunities. Fluctuating power prices and community desire for new renewables also play important roles in motivating community mobilization. The data on these trends is presented below.

Declining Distribution Co-operatives

Distribution co-ops represent a declining share of the electricity co-operatives in Canada; generation retail and education co-operatives are growing in number. Table 6.3 illustrates the diversity of electricity co-operatives across Canadian provinces by date incorporated. Distribution co-operatives in this period are heavily concentrated in Alberta (89 percent); generation co-operatives exist in all provinces but are concentrated in Ontario and Quebec (71 percent and 16 percent respectively); consumer and networking co-operatives are concentrated in Ontario.

The distribution co-operatives remaining in Alberta today are divided between seven that own *and* maintain the grid (self-operating distribution co-ops) and fifty-four that own the lines but contract out to IOUs like Fortis

TABLE 6.3
Electricity co-operatives by province and type incorporated, 1990–2013

Type	BC	AB	SK	MB	ON	QC	NB	NS	PEI	Canada
Distribution	-	17	-	-	1	-	-	1	-	19
Generation	1	1	2	3	91	21	6	3	1	129
Consumer	1	1	-	-	3	-	2	-	-	7
Networking	-	1	-	-	10	-	-	-	-	11

Sources: CCA 2011a; Co-operatives Secretariat Survey 2010; provincial associations and project websites.

and ATCO Electric to manage their assets. In 2011, approximately 20,000 (or 50 percent of rural electrification association, or REA, members in Alberta) were served by one of the seven self-operating REAs, and the remainder are members of the more than fifty-four small ones operated by IOUs (personal interview, November 27, 2009; CAREA 2012). These co-operatives are diminishing in number as assets are sold to the province's private utilities and some are amalgamated with other co-operatives. According to the Alberta Federation of Rural Electrification Co-operatives, 143 amalgamations and mergers plus 255 sales have occurred in the province, of the original 398 incorporations since 1940 (Nagel personal communication, January 2011). These changes are due to various reasons, explored below, chief of which is pressure from private utilities in the province's deregulated energy market.

According to a February 2012 press release from the Central Alberta Rural Electrification Association (CAREA), Lakeland REA, and South Alta REA, "the rules in Alberta are working against the REAs because of the increased investment levels and the profit-maximizing investment of the multinational utilities ... The ability of the multinational utility companies to raise their investment levels in the past couple of years now effectively allows them to also utilize these levels to purchase REAs and rewards them for doing so" (CAREA 2012). This pressure has resulted in sales of co-operative distribution infrastructure to either ATCO Electric or Fortis, the two private power distribution companies in the province (Bourne personal interview, December 1, 2009). According to one interviewee from the Rural Utilities division of Alberta Agriculture and Rural Development, deregulation has played a role in this:

> The electricity market since 1995 has been in a state of turmoil under deregulation, and this has a significant impact on REAs and how they're

dealt with. They have to live with the rules of deregulation ... The issue that's facing the REAs is there is still a desire by ATCO and Fortis to acquire the REAs, and they'll snap them up as soon as there's any interest in doing this, to the detriment of other REAs, like CAREA, which wants to have one REA in the whole province. They are trying to minimize the obstacles to their amalgamation with others and to increase obstacles to prevent sale to other utility companies. (Rural Utilities employee, Alberta Agriculture and Rural Development, personal interview, November 27, 2009)

The president of the Alberta Federation of Rural Electrification Associations (AFREA), Al Nagel, pointed out that sales also occur because of a lack of co-operative awareness and short-term economic interests of a membership pressed for cash on other fronts:

The whole age thing is important for someone who has been on board for a hundred years. They're tired, and deregulation has made things twice as complicated with all these new forms to fill out and now you've got to have a computer. If you just look at it from a financial point of view, we're in the co-op and we farm for fifteen or twenty years and someone offers you $20,000 and your rates will remain the same, what do you say? Especially in today's economy where crops aren't good, cattle prices aren't good. Even though I'm going to have to pay it back through my rates eventually, and my successors will forever. (Nagel personal interview, November 27, 2009)

The result of the less hands-on (and more common) model of REA is that many co-op members are unaware of the co-op functions, benefits, or purpose. So, when an offer to buy the local lines arises, with a one-time payment to each member of $5,000 to $10,000, members sell. To combat this trend, some REAs have elected to amalgamate with each other. The ones that have not folded have had to spend large sums of money. According to a board member of both CAREA and an Alberta gas co-operative:

The REAs don't have controlled franchise areas. So we're constantly under attack and have spent huge resources defending ourselves against them ... they just keep picking at you and picking at you and picking at you and one day they might find a crack and they'll game it. But the resources, when I talk about saving our members $750 each a year, that could have been $1,750 had we not been spending all this money defending ourselves ... We go to arbitration with them constantly, [there's] never a time that we don't have one going on, sometimes two or three at a time. If we were able to turn the clock back, the REAs would die to have a controlled franchise area where they owned all the distribution rights like the gas co-ops. The difference is staggering. I was shocked to sit on both boards. The government didn't necessarily set up the REAs properly; there was more vision with gas co-ops, [and they] made corrections for mistakes in REAs. (Personal interview, December 1, 2009)

Quebec and Alberta are not the only provinces with power distribution co-operatives. Ontario (Embrun) and Nova Scotia (Municipal Electric Utilities of Nova Scotia Co-operative) also have just one distribution co-operative each, both incorporated after 1980. In each of these cases, earlier municipal power systems were the precursor to forming the co-operative, unlike the broader development of co-operatives in Quebec and Alberta. Embrun, for example, was a local distribution company before the restructuring of Ontario Hydro and chose to go the co-operative route, unlike other local distribution companies in the province.

The Rise of Renewable Generation Co-operatives

Inspired by the success of German and Danish co-operative and community power development, Canadian co-operative electricity-generation projects are now developing in ever-increasing numbers (Gipe phone interview, April 7, 2010; personal interview, July 23, 2009; McLean personal interview, July 23, 2009). They are facilitated by electricity-sector restructuring that relies on private IPPs for new generation. In British Columbia, government-mandated IPP development has tripled the share of private generation, from 3 percent in 1999 to almost 10 percent in 2011 (Statistics Canada 2014). As private corporate actors, Canadian Utilities, AltaGas, and Spain's Acciona Energy among them, develop projects from British Columbia to Nova Scotia, some communities, particularly those with a successful co-op model to point to, are keen to participate. Presented below is a brief overview of generation projects that involve co-operatives across Canada. Chapter 7 examines their project structures and potential in more detail.

Of the at least 166 electricity co-operatives incorporated in the 1990–2013 period, nearly 80 percent are focused on power generation. These include both co-operatives (e.g., WindShare) and co-operative-initiated projects (e.g., Bear Mountain). Unlike earlier developments of electricity co-operatives in Canada, they are concentrated in central and eastern Canada, most notably Ontario. Nationwide, of the roughly 132 co-operative electricity-generation projects incorporated since 1940, at least 20 are generating electricity. In the most recent Federation of Community Power Co-ops survey (FCPC 2015, 18), their membership reported having 67 FIT or microFIT projects currently generating power.[9] The vast majority are not generating electricity as yet, even if the co-operative was incorporated for this purpose. In Ontario, for example, with an active FIT program in place since 2009, only 4 co-operatives were generating power in 2013 (FCPC 2013, 2014).

The relatively small number of co-ops actually generating or involved in the generation of power is changing, as contracts awarded in recent years are starting to develop. These include 59 MW of installed capacity for wind co-operatives in Ontario and Quebec in 2010 (total wind installed capacity in Canada, by comparison, was 4,008 MW in February 2011). In Ontario, a further 130 contracts were issued to co-operative (fully or partially owned) wind and solar projects in 2013 and 2014. If all are installed, this will result in an additional 27 MW of co-op power (FCPC 2013). However, even with these contracts it takes many years for a project to be fully financed, built, and grid connected.

Although this is still a very small change in terms of the overall profile of power generation in Canada, it represents a significant shift from co-operative participation in this area just thirty years ago. The focus of new projects has shifted in recent years from wind turbines, which dominated early RESOP/ FIT proposals, to solar projects. According to a co-operative developer, "solar is just easier. It is not just the [higher] rates ... the FIT is designed to provide a rate of return. It is just easier from regulatory and construction points of view. Wind can be pretty complex and almost always controversial" (personal communication, September 2013).

Consumer, Worker, and Networking Power Co-operatives

Consumer and networking co-operatives in the electricity sector represent a minority (eighteen) of the new and existing co-ops. This number is likely understated because many of the generation co-operatives that have started act as networking and advocacy organizations while mired in the project-development stage (and many fail to move out of this stage). In 2013, five networking power co-operatives were on record, nearly all in the province of Ontario, and seven electricity consumer co-operatives were spread across the country. Nonprofit community renewable co-operatives focus on conducting educational campaigns for sustainable and renewable energy, and sometimes, as in the case of the Toronto Renewable Energy Co-operative, act as an incubator for generation co-operative project spinoffs. These co-operatives also play an important role in networking and outreach for the broader community power sector (see Chapter 8).

Consumer co-operatives in the electricity sector have bulk purchase power and power technologies in restructured power markets. Co-operative members can ensure that they pay closer to the wholesale price for power and, through member control of the co-op, can contract for power generated by renewables. As of early 2015, these consumer electricity retail co-operatives exist only in Alberta and Ontario. These two provinces have had the wildest price fluctuations for electricity and are the only two provinces in Canada with retail markets for electricity at the household level. In Ontario, Canada's largest farm energy co-operative runs FireFly Energy, which acts as an electricity retailer to co-op members.

Alberta's Spark Energy Co-operative is also a power retailer. Members buy shares and purchase their power through the co-op. It then uses the funds to buy wind, solar, and biomass electricity from Alberta's Power Pool. Renewable-energy certificate systems like this are plentiful in Alberta. They facilitate wind development in the province by paying higher than market price for green power sources. This creates a market for green power in that province that may make new projects more financially viable. Self-operating REAs in Alberta can also act as power retailers. The largest REA, Central Alberta Rural Electrification Association (CAREA), started a green-tags initiative so member-owners can, for a supplementary fee, purchase renewable electricity. For $20 a month, members of the REA can purchase 1 MWh of renewable energy that is "physically metered and verified in Alberta" (CAREA 2011).

Finally, co-operatives exhibit potential to more affordably retail goods for energy efficiency and conservation. Although most observers concur that voluntary and individualistic measures toward conservation are not nearly sufficient to reduce demand, co-operative home audits, installation, and consumer initiatives can play a role in making these actions more affordable for a wider swathe of the Canadian population. Furthermore, a number of these (seven so far) are incorporated as worker co-operatives. BC-based Vancouver Renewable Energy Co-operative and Viridian Energy Co-operative install and provide consulting services for local energy efficiency and home-based generation. The Sustainability Solutions Worker Co-op and Fourth Pig Worker Co-op also are structured so that meaningful employment is provided, as well as services for energy sustainability. Despite the small number of these consumer, worker, and retail co-operatives to date, it is reasonable to expect these types to play a larger role in the co-operative electricity sector in years going forward as both conservation and power trading gain importance.

Policy Supports for Renewable Generation Co-operatives

Public policy and politics mattered for electricity co-operative development in the past, just as it does today. The popular idea that co-operatives are independent, apolitical, bottom-up institutions (see Chapter 3) is simplistic. Electricity generation co-operatives began to develop in Canadian provinces *after* the private sector began developing new renewables. However, a small number of pioneering co-operatives developed before policy supports were put in place for them and were instrumental in pushing for policy changes (see Chapter 8). This was the case for the Toronto Renewable Energy Co-operative's WindShare project. Policy supports then may raise the number and the profile of these actors, allowing them in some cases to scale up. Provincial government support for community power development can increase acceptance of new renewables development and also of neoliberal policies directing public funding to private actors.

The majority of new generation co-operatives are concentrated in provinces with the most open electricity sectors and those with community power policies (e.g., Ontario). All of these policies include co-operatives in the definition of "community" but are not focused solely on co-operatives. New community policy initiatives were initiated in Ontario, Quebec, Nova Scotia, and New Brunswick over the past seven years (see Table 6.4). There are three mechanisms being used by the provinces: community feed-in tariffs (COMFITs), set-asides, and start-up funds. A COMFIT guarantees a fixed price for power to generators that meet the definition of "community" set out in the policy regulations. This price is based on the cost of generation, plus a "reasonable return." If a project qualifies for the COMFIT, it is awarded a long-term contract with the power authority. A set-aside is a mechanism whereby a portion of new power coming onto the grid is reserved for community actors. For example, if the central power authority is set to increase generation by 15 percent, it might sign contracts for only 10 percent of that, leaving the final 5 percent for community groups. Finally, some of these policies are providing start-up funds for community-based projects to help with the costs of feasibility studies, contract and project design, and in some cases construction.

TABLE 6.4

Provincial community power policies, 2014

Province	Year	Policy	Prices	Limit/set-aside	Notes
ON	2010 FIT 1.0 2012 FIT 2.0 2014 FIT 3.0 2015 FIT 4.0	FIT with community and First Nations	+1¢/kWh for community; +1.5¢/kWh for First Nations	21 MW for majority community projects in FIT 3.0	FIT prices differ for all generators based on project size and source Projects < 500 kW
QC	2009–13	Community windpower calls	Call for tender (bid) 2013 6¢/ kWh 2009 call average, 13.3¢/kWh	All 450 MW of 2013 call must have community participation	Project in 2009 call limited to < 25 MW
NB	2010	Modified COMFIT	10¢/kWh for community renewable energy projects	75 MW in total (50 community, 25 First Nations)	Projects < 15 MW
NS	2010, ended 2015	COMFIT	13.1¢/kWh wind > 50 kW 49.9¢/kWh wind < 50 kW 65.2¢/kWh tidal	200 MW in total	Prices differ based on project size and source Projects < 500 kW (2014 review)

Sources: New Brunswick 2011; Nova Scotia Department of Energy 2012, 2014; OPA 2011, 2012, 2014; Québec 2011, 2015.

Ontario: Pioneering Feed-in Tariff Adders and Funds

In North America, Ontario has gone the furthest to support the development of community and co-operative generation projects. The Ontario government's 2009 Green Energy Act provided two things that have stimulated more co-operative generation development in that province: (1) a feed-in tariff, a renewable energy procurement policy that some argue (e.g., Gipe 2007, n.d.; Lipp 2008) favours community-based projects and (2) an adder for community (co-operative) and First Nations generation projects.[10] Of particular note was the initially very high (80 cents, reduced in FIT 3.0 to 39 cents) feed-in tariff for solar projects. Specifically targeted modifications to provincial co-operative legislation made it easier for renewable energy co-ops to incorporate in Ontario.[11] According to the Federation of Community Power Co-ops, the Ontario government has supported community power in the following ways:

- Introduced the Green Energy and Economy Act (GEA);
- Established a Feed-in Tariff (FIT) program allowing community power proponents to apply for power purchase contracts;
- Amended the Co-operative Corporations Act (Co-op Act), allowing renewable energy co-ops to engage member-investors more effectively;
- Established community power and Aboriginal grant programs to support early stage soft costs for projects;
- Established price adders for community power and Aboriginal groups;
- Established a $1 million annual set aside of grant money to build capacity in the community power sector;
- Established set-asides for community power co-ops and Aboriginal groups, and more recently municipalities, under the small FIT Program (~135 MW by 2016); and,
- Established loan guarantees for Aboriginal groups. (FCPC 2013, 4)

The community power-policy element of the Green Energy Act described in Chapter 5, as with the act itself, was the result of years of policy pressure and lobbying by community actors in that province as well as of interest from government leaders (the premier and energy minister) in that province. Policy learning from Ontario is now spilling over into other jurisdictions: "Ontario ... had a broad coalition that consisted of agricultural groups, the biggest unions (steelworkers), church organizations, environmental organizations, and business leaders. This turned into perfect storm with a change in ministers in the energy department; all of the sudden, we had an engaged minister who saw an opportunity" (Ontario co-operative developer interview, 2010).

However, in FIT 1.0 (2009–11), only 7 percent (329 MW) of contracted power was allocated to communities (including two co-operatives) and 10 percent (479 MW) to First Nations (OPA 2011). FIT 2.0 rules were amended in 2012 to provide new contract capacity set-asides (CCSAs) of up to 25 MW for co-operatives and another 25 MW for First Nations groups (see Chapter 5). For communities to qualify, "the project must show evidence that the co-op has a direct economic interest of at least 50% in the project. In addition, the applicant must provide consents and declarations in the prescribed form from at least 50 members of the co-op that are property owners in the host municipality" (OPA 2013a). The results in July 2013 were promising: thirty-one co-operatives were awarded FIT contracts, sixteen of which had a majority interest in the project. However, many more co-operatives applied than were awarded contracts (FCPC 2013, 2014; OPA 2013b).

Under FIT 3.0, projects that can demonstrate community, Aboriginal, or municipal participation of more than 50 percent are still eligible for CCSAs. FIT 3.0 was announced in July 2014. The Ontario Power Authority received 1,982 applications for a total of 493 MW and offered contracts for 405 of them (123 MW). Nearly all of these were capacity set-aside (majority-owned) projects for municipalities, First Nations, and co-operatives. The share of contracts with community participation also increased substantially, with 51.4 percent (60 MW) Aboriginal participation, 16 percent (21.5 MW) community participation, and 32 percent (42 MW) municipal or public-sector entity participation. As noted in Chapter 5, however, eligible FIT projects were capped at 500 kW, so large developers were unlikely to apply for this program; most co-operative wind developments would be ineligible now given current commercially viable turbine sizes. One project developer, Reliant First Nation Limited Partnership (with Adelaide Solar Energy and Solar Income Fund), was awarded 31 percent of the total MW. Six solar co-operatives and one biomass co-operative were awarded contracts for 7.56 MW (6.1 percent). More than 98 percent of FIT 3.0 contracts were awarded for solar projects.

In addition to the financial incentive the FIT adders provide for community power, and the new priority system for small community projects, until early 2015 Ontario also financed soft development costs. As of November 2014, the Community Energy Partnerships Program was taken over by Ontario's IESO and now includes the Municipal and Public Sector Program, Aboriginal Energy Partnerships Program, and the Education and Capacity Building Program. It provides grant support of up to $10 million annually, but has put this on hold while working through a consolidation of programs and agencies in Ontario's electricity sector. Prior to these changes, the Community Power Fund, a nonprofit co-operative, administered three programs: the Community Power (CP) Fund grant program, the Community Energy Partnerships Program, and Community Power Capital. They received on average 31 applications per month, and 308 since the Community Power

Fund's inception up to 2013. Co-operatives represented 8 percent of these applications, whereas farmers represented 40.6 percent. More than $8 million in funding has been granted through the program to solar, wind, hydro, biomass, and biogas project development (Green Energy Act Alliance 2011).[12] No other province in Canada had this range or amount of financial support for community power development, but as discussed in the previous chapter, recent changes to the programs have made further expansion challenging.

Quebec: Community Power Calls

In Quebec, community power-policy supports arose after Hydro-Québec started issuing power calls for private wind development; its community policy specifically targets wind generation. The first general wind-power call was in 2005, with subsequent private procurement calls in 2008, 2010, and 2013. Together they total 4,000 MW of installed capacity (Québec 2011, 2015). As a result of these new opportunities, co-op developers worked to create a dozen electricity co-ops in the Bas-Saint-Laurent region, but were not awarded contracts in the competitive bidding system. In 2010, Hydro-Québec made a third call for power for an additional 500 MW to be split between community (250 MW) and First Nations (250 MW) projects. In the November 2013 call for 450 MW, all projects needed at least 50 percent "local environment" participation, which included a regional county municipality, a local municipality, a Native community, an intermunicipal board, and a co-operative where the majority of members live locally. Although in Ontario projects can have 15 percent community investment (with a reduced adder), in Quebec, Nova Scotia, and New Brunswick, they need at least 50 percent to qualify.

The procurement mechanism for Quebec's power calls is lowest bid, and according to a co-operative researcher in Quebec, "this process favours the big companies and multinational corporations. In this context, it's very difficult for the development of energy co-ops in Quebec. The government decided to reduce social opposition to their power policies. Community power will be residual, 250 MW. It's not big, but it is an overture" (personal interview, May 13, 2010). Funding supports for project development are not included in the Quebec policy, but Quebec has a widely recognized program of co-operative development system in the province (Laville, Levesque, and Mendell 2007).[13]

The results of the first community-focused power call were announced in December 2010. In all, 291.4 MW of power was allocated in the contracts selected, out of a possible 500 MW, to be constructed between 2013 and 2016. Only one of the twelve bids was a co-operative project, Val-Éo (24 MW), and only one was an Aboriginal project (also 24 MW). The other ten community projects were put forward by municipalities, often in partnership with private companies – Innergex, Saint-Laurent Énergies, Algonquin Power and Utilities, Boralex, and Northland Power (Hydro-Québec 2010). Eighteen windfarms are currently operational (2014) in Quebec, with an installed capacity of 2,187 MW and with over 1,100 more MW planned or under construction (Hydro-Québec 2014). The community and First Nations content (with nearly half of this content coming from private partners) represented just a tiny portion of Quebec's wind-power calls, despite the strength of the co-operative sector in Quebec. Only one co-operative has secured a power contract and this was in partnership with Algonquin Power and Utilities.

Nova Scotia: Feed-in Tariffs, Set-asides, and Community Investment Funds

In 2010, Nova Scotia announced a community feed-in tariff (COMFIT) following a range of consultations with Nova Scotia Sustainable Energy Association (NovSEA) members and industrial stakeholders. This policy emerged out of the failure of earlier renewables policies to diversify ownership outside Nova Scotia Power. Before 2011, the Nova Scotia government mandated that the private Nova Scotia Power purchase new renewables (wind, solar, tidal) from independent power producers (IPPs). However, because it was through an RFP (request for proposal) system, which led to lowest-bid initiatives, projects went bankrupt, due to underbidding and the economic crunch that hit in 2008/2009: "Nearly all the IPPs failed. Because of that, the government allowed Nova Scotia Power to buy them out and finance up to four years some of them (and still consider them IPPs), so that rule really wasn't abided by, otherwise Nova Scotia Power wouldn't have met the 2010 target" (Ashworth personal interview, May 21, 2010).

The Nova Scotia COMFIT program started accepting applications in September 2011 once the FIT rates were set. As of December 2015, ninety-three COMFIT projects with a total capacity of 200 MW were approved (Nova Scotia Department of Energy 2014; Nova Scotia Government 2015a). Contracts were awarded for projects ranging from a 50 kW turbine in Pictou to a 6 MW combined heat and power biomass plant owned by Cape Breton Explorations. To be eligible for COMFIT, the projects had to be community owned and connected at the distribution level (typically under 6 MW). The COMFIT is part of a larger renewables procurement plan mandated by the province. Six hundred megawatts were targeted through competitive bidding; community actors could apply for a hundred MW of contracts under the COMFIT program in 2011. As with the FIT in Ontario, the rates varied based on the source of power generated, ranging from 65.2 cents per kWh for small-scale tidal to 13.1 cents per kWh for wind projects greater than 50 kW. Unlike Ontario's FIT, however, only community groups could access the FIT contracts, and these projects were limited to the distribution system (capped at around 2 MW in practice, depending on the local grid capacity). Partnerships were allowed as long as the community partner owned 51 percent of the project.

The following excerpt from e-mail correspondence with Nova Scotia's former energy minister (2009–11), Bill Estabrooks, illustrates the intent of the COMFIT:

> To support these projects, government will establish a sustainable energy planning group to assist new developers with business plans, technical feasibility studies, grant applications, public outreach, regulatory approvals, and financing guidance. We are also working on developing appropriate finance tools for community-based projects. We have also been involved in a unique pilot project involving wind energy development in municipalities. Funding from the province will allow municipalities to work with residents to identify locations in their communities where wind energy development is encouraged, discouraged, or prohibited. The funding is also intended to increase public involvement and educate Nova Scotians about renewable energy and environmental goals. (Estabrooks personal communication, May 6, 2010)

The Nova Scotia case is unique in that an innovative community investment mechanism, Community Economic Development Investment Funds (CEDIFs), also qualified under the "community power" definition. These groups were very successful under the COMFIT, along with First Nations and municipal organizations; to date, no co-operatives have been awarded a contract. CEDIFs are far more numerous than co-operatives. One group in particular, Scotian Windfields, sees itself as going to fill most of the 100 MW (75 percent) (Zwicker personal interview, May 20, 2010). It was also involved in pre-COMFIT IPP development. According to CEO Barry Zwicker:

> We were able to win an RFP for a 30 MW project in Digby, an $84 million project. We had a partner with us at the time, SkyPower, and they were basically a Lehman Brothers company. Lehman Brothers went bankrupt,

so SkyPower went bankrupt, which put us in a very difficult position, and we ended up selling to Emera [parent company of NS Power]. (Zwicker personal interview, May 20, 2010)

The first COMFIT project came online in March 2013, 300 kW of wind installed capacity built by Halifax's Seaforth Energy and owned by the municipalities of Goldboro, Tatamagouche, and New Glasgow (Nova Scotia Government 2013). Many of the projects have yet to reach the fully operational stage. Furthermore, the Nova Scotia government limited applications for community wind projects to 500 kW in 2014, citing grid capacity and resource integration issues. The report from the 2012 COMFIT Review argued that

> large-scale applications generally have limited community equity contributed just meeting the program requirement of 20 percent. Very few applications have equity contributions more than 50 percent, or are 100 percent community owned (excluding biomass). Increasingly large scale applications (2MW+) are commercial operations where private partnerships are maximized not community investment. This is beyond the scope of what the FIT rate was originally intended ... The province is nearing the technical limit for intermittent sources of energy and it is yet uncertain how many COMFIT projects will ultimately come online ... continued acceptance of large scale applications (2–4MW) may begin to interfere with the transmission system as we approach 500MW of intermittent sources of energy. (Nova Scotia Department of Energy 2014, 12)

Following the October 2013 election of a Liberal government and a broad electricity sector and COMFIT program review, in 2015 Nova Scotia's Energy Minister announced that the COMFIT program would not be accepting any new applications. The government cited as the main reasons that the program had exceeded the desired energy output, excessive project development times, and a lack of will by the public to continue funding it. The Department of Energy announcement pointed out that "evidence shows us that COMFIT renewable electricity represents 15 per cent of Nova Scotia Power's fuel cost but only five per cent of the electricity generated. Nova Scotians have been clear with us about their support for greener sources of energy, but not at any cost" (Nova Scotia Government 2015b).

New Brunswick: Modified Community Feed-In Tariffs
The New Brunswick Community Energy Policy (CEP), announced in 2010, differed significantly from those in Ontario, Quebec, and Nova Scotia.

It was the result of a round of consultations beginning in 2008, and is a modified COMFIT. Unlike in Ontario and Nova Scotia, there was no differentiation in the price based on generation source. All qualifying projects would receive 10 cents per kWh (New Brunswick 2010). This was seen by many observers as too low to provide sufficient return to get projects constructed. The FIT rate was frozen for the first five years, then set to rise with the consumer price index in New Brunswick. One interviewee familiar with the progress of the policy suggested that it was initiated to see what kind of interest there was and not actually expected to generate projects: a first-step test, to balance cost considerations with new renewable development.

Round one of the 2010 CEP consisted of a call for 75 MW: 50 for community and 25 to First Nations projects. These were capped at a maximum project size of 15 MW and, again, partnerships are allowed with private corporations as long as a municipality, co-operative, First Nation, or nonprofit is the majority shareholder. Twelve communities responded to the call for interested parties, despite the lower FIT rates for wind than in both Nova Scotia and Ontario.

The Liberal government in New Brunswick lost the 2010 election to the Conservatives, in part because of popular anger over its proposal to sell the province's power utility (Moore 2010).[14] The Liberals then regained power in 2014. During the interim (2010–14), no projects were announced under the CEP, but following a 2015 review the New Brunswick Department of Energy and Mines announced a new community energy procurement program, Local Renewable Energy Projects That Are Small Scale (LORESS). It is for 80 MW in total, with individual project sizes of up to 20 MW each and no feed-in tariff. The government cited cost considerations when justifying the choice to cancel the COMFIT (New Brunswick n.d.). A call for 40 MW of First Nations project proposals was announced in January 2016 and a community tranche is expected in 2017.

Summary

Co-operatives are developing today in response to concrete environmental and economic development challenges, with mixed success. They are taking on technologically complex capital projects and attempting to compete for sites and contracts with some of the largest energy companies in the world. They are also moving from early concentration in distribution in Alberta to a more even regional distribution between eastern and western provinces and between generation, retail, education, and distribution. Canadian rural electrification associations have made a very real contribution to the life and development of Albertan communities over the past seven decades. Electricity co-ops emerged when energy resources were prohibitively expensive and communities were forced to innovate to survive.

Electricity co-operatives in recent years have developed in various new ways across Canada. Public policies have played a crucial role in facilitating new developments, through both opting for private-sector ownership and being slow to take on climate change and new renewable-generation sources. New policy supports include set-asides and power calls specifically for community power, funding for community projects in the form of seed grants, feed-in tariffs for community projects only, and a range of other initiatives. The historical record demonstrates, however, that ideology matters to co-op development; co-operatives have historically been used by governments to provide services they are ideologically opposed to funding outright. Contemporary funding pressures and political challenges across the provinces are prompting a reconsideration of community and renewable energy policies alike, as the full costs of system restructuring start to be tallied. These lessons are important for our understanding of how and where co-operatives may contribute in the future.

7

Off the Ground and on the Grid

New Electricity Co-operative Development

New co-operatives are developing in electricity generation across the country, with significant variation in how they are owned and structured, and in the sources of power they generate. Despite newly enacted policy supports in some provinces, however, significant challenges are faced by electricity co-operatives today. On the one hand, the shift toward private generation of power for new renewables and the community power policies examined in earlier chapters can create space for them. This is especially evident in Ontario. On the other hand, the structure and power of actors within the electricity sector place co-operatives at a disadvantage to other developers and the policy context for community targeted support can change suddenly when new governments step in. Sectoral competition for sites, financing, and grid access have challenged project proponents and resulted in a wide range of new generation co-operatives in terms of size, ownership, and democratic constitution. These new co-operatives are forming accompanied by great enthusiasm over the prospects for local development and community power but are often unable to move to completion or, if they do, they do so with a minimal portion of project ownership. Despite the promise of their relatively democratic constitutions and a capacity to engage in innovative renewables projects, key issues of financing and co-optation remain.

The co-operative difference plays out for these new co-operatives in distinct ways, depending on the scope and scale of the challenges confronting them. By examining specific projects, we can understand more fully how electricity co-operatives have contributed to community ownership and control, and how they may set a foundation for building structures of empowered participatory governance. For example, co-operative generation projects have helped develop new renewables in solar and wind power in a way that minimizes local opposition to projects and keeps at least a portion of the project profits in the community. In addition, retail and consumer electricity co-ops are well placed to take the co-operative strengths in bulk purchase and installation projects to support more affordable and efficient retrofits and microgeneration, extending these options to a broader segment of the population. Finally, beyond the actual physical ownership of generation and distribution of electricity, co-operatives are playing an important role in new renewables education.

The challenges that have emerged for new generation co-operatives include site access, regulatory approval, and – the most pernicious challenge in the post-2008 economy – financing. When markets are created for private power development, co-operatives secure only a very small portion of new contracts, if that. According to many members and policy makers interviewed for this book, this is because without prior project-development experience, deep pockets, and significant time and energy, or even all three together, securing loans can present an insurmountable challenge (personal interview, July 20, 2009; Mole phone interview, July 20, 2010; Zwicker personal interview, May 20, 2010). Community-based actors rarely have any of these, working, as they do, with member financing, government grants, and a significant amount of sweat equity.

These varied challenges result in private partnerships with a larger entity – either a municipality (as was the case with Canada's first urban co-operative wind turbine, WindShare) or a private developer (as with the Bear Mountain project). Depending on the actor, these partnerships may significantly reduce the depth of the social economy and sustainability benefits of the project, in terms of both control and the local multiplier effect. So, partnerships are accompanied by advantages (funding, experience, and risk-sharing) as well as by the drawbacks associated with leveraging community support for minimal share and project control. Some partnerships serve to significantly weaken the contribution of co-operative electricity projects to economic democracy and empowered participatory governance. Sectoral competition presents another challenge. The growing pressure on distribution co-ops in Alberta – the province where they're the most numerous – to sell and fold is one example of this, as is the need most projects have to find private-sector partners.

Promises: Local Ownership, Participation, and Education

Electricity co-operatives can differ from other power-sector actors on numerous fronts. They can embody local ownership, participatory governance structures, and objectives that transcend profit as a sole motive (referred to in the literature as "multiple bottom lines"). Because of these differences, electricity co-operatives are accompanied by a range of potential local benefits once they are up and running, including but not limited to controlling resource allocation, providing local jobs, ensuring lower prices (in some cases), and delivering more responsive service. According to a 2011 report by the Green Energy Act Alliance,

> the Pembina Institute for the Community Power Fund modeled community job impact for community energy projects. The findings of this study are congruent with existing literature, which suggests that community-owned power projects lead to more local jobs than traditional development of similar projects. A literature review also suggested that additional benefits, such as increases in project participation and project acceptance and a decrease in project resistance, can also result from community power projects. (Green Energy Act Alliance 2011)

Co-operative power development also plays a role in reducing social opposition to the development of renewables (Musall and Kuik 2011). This reduced NIMBYism is favoured by environmental advocates looking to move toward cleaner generation sources (Emond 2010) and by policy makers looking to legitimize power-sector investments and changes (Nova Scotia Power personal interview, May 19, 2010).

Co-operatives today can also be differentiated by the stage of development they are in (operational, in progress, and inactive). Co-operatives in the project-development stage, for example, are involved in volunteer board meetings and project planning, developing the financing packages, raising capital and setting up, for instance, wind monitoring towers, whereas inactive and stalled co-operatives are those that have ceased meetings and sold or demutualized the co-operative. It is likely that many co-operatives never reach the incorporation phase, so information on them is less readily available for analysis. They fail to establish a web presence or to register with local co-op associations. Table 7.1 illustrates the various stages of development that electricity co-operatives in 2013 were in.[1]

TABLE 7.1

Status of electricity co-operatives in Canada, 2013

	Operational	In progress	Inactive/stalled/sold

New electricity co-operatives	2 electricity retail
	6 installation
	2 hydro generation[1]
	3 wind generation
	5 solar generation
	2 bioenergy
	10 education/
	  networking
	1 finance
		14 generation
		  projects with
		  contracts awarded
		  (10 solar, 1 biogas,
		  3 wind)
		27 wind generation
		51 unspecified
		12 solar generation
		6 biomass generation
			11 wind generation
			2 hydro generation
			1 solar generation
			2 unspecified
			2 electricity retail

Subtotal	31	110	18
Existing distribution	51 distribution REAs 10 self-operating REAs		255 REAs sold to IOUs,[2] 40 to Hydro-Québec
Total	92	110	313

1   Irrigation canal, a co-operative or municipal irrigation district in Alberta.
2   This does not include amalgamations in Alberta.

Asset Ownership and Local Investment

Electricity co-operatives take a wide range of ownership forms. In order for local alternative projects to amount to more than participatory window dressing and to have significant impacts on electricity governance and environmental sustainability, they need to move from consultation toward a high degree of countervailing power. This, together with participatory collaboration, is what forms empowered participatory governance. Countervailing power requires an organizational network, mobilization, and a resource base. Importantly, some electricity co-operatives own and manage the projects they initiate, whereas others do not. Below I highlight some of the co-operative projects that are operational (or very close to it) and demonstrate a range of project forms. In some cases, such as the WindShare Co-operative turbine in Toronto, the level of member participation was high, the ownership was local, and the co-op formed the basis of what is now a growing community power sector in Ontario. This project also included a significant component of co-operative education that reached down to the membership and out to the community at large.

The 750 kW WindShare turbine at Toronto's Exhibition Place is the first urban 100 percent community (municipal and co-operative) wind-generation project in North America.[2] These urban wind pioneers have made a significant impact on community energy beyond the city of Toronto. The project was initiated by the Toronto Renewable Energy Co-operative in 1999 and completed in 2002. The WindShare turbine is a 50-50 joint venture between a municipal power utility – Toronto Hydro (Energy Services Inc.) – and WindShare Co-operative. WindShare has over six hundred co-op members, 99 percent of whom are in Toronto. Minimum investment was $500 per member, and the average investment in the project was between $1,000 and $2,000. According to the former president, Evan Ferrari, new community members wanted to join the project even when it was fully subscribed, so $250,000 was allocated to a trust account to be put toward future projects. The total cost of construction and installation of the turbine was $1.6 million, with $800,000 of this invested by the co-operative. Today, the project generates enough electricity to power two hundred homes (Ferrari personal interview, July 23, 2009). Any revenues from the project circulate back to members through dividends set by the board and approved by the membership.

WindShare Co-operative's membership is drawn from residents of the city of Toronto. By contrast, the Val-Éo Co-operative in Quebec is a solidarity co-operative, its membership drawn from sixty farmer-landowners, employees, local residents, and two municipalities. The co-operative was launched in 2005 when local farmers started receiving proposals from private wind developers to lease their land. They joined together to form the co-operative, erected the meteorological towers to study the local wind resource, and from there negotiated a partnership with Algonquin Power and Utilities to develop a 50 MW wind farm on member lands, enough to power approximately 13,500 homes. One of the innovations of this project was that co-operative landowner members could decide to contribute land instead of cash to the project, still receiving membership shares in return. The project failed to win a bid in Hydro-Québec's 2007 call for tender but was ultimately successful in winning a power purchase agreement under Hydro-Québec's 2010 community power call. Construction on the project is set for 2016.

Projects wholly owned by co-operatives are more common in new solar developments in Ontario than in wind generation (see Table 7.2). In fact, with retail, educational co-ops, and self-operating REAs, partnerships for project ownership and control haven't emerged. These differences are likely because the higher the capital investments required for wind and hydro generation projects, the more need there is for private partners. In the case of the solar projects – SolarShare, Ag Solar, and AGRIS Solar – the solar feed-in tariff rates under the Green Energy Act FIT 1.0 were relatively high, ranging from 71.3 cents per kWh to 44.3 (compared with 13.5 cents for wind) for FIT, and 80.2 to 64.2 cents per kWh for small microFIT projects (AGRIS Solar Co-operative 2012; OPA 2010). FIT projects are also eligible for a community or First Nations adder of between 1 and 1.5 cents per kWh.

SolarShare, another project initiated by the Toronto Renewable Energy Co-operative, is a community investment co-operative wherein members purchase community solar bonds for $1,000 (plus a one-time $40 membership fee) and in return receive a 5 percent annual return for five years. When all awarded FIT projects are developed, it will be the largest renewable energy co-operative in North America, with close to 8 MW of power capacity. Co-operative members vote to elect their board, and the co-op provides resources if they wish to initiate projects in their local community. The co-operative had built thirty-three projects as of December 2015. Sunfield (built in 2011) is a 195 kW array – a connection of numerous individual solar cells – that cost $1.7 million and generates 320 MWh of power per year and $257,000 in annual revenue. Waterview was the largest community-financed rooftop project in North America in 2011, a 438 kW rooftop solar system that cost $2.27 million to build and returns $323,000 per year to the community investors (SolarShare Co-operative 2015). Its other projects are spread across Ontario – Toronto, Ottawa, Moose Creek, and St. Catharines – bringing total current capacity to more than 5.5 MW.

AGRIS Solar, in Ontario, is a farmer-based solar co-operative with over seven hundred members. The co-operative installs and maintains solar panels on member property, reducing the cost per unit and aggregating other costs such as insurance risk. Members are paid a licence fee for use of their property and receive a share of the co-operative profits. In 2014, the co-operative had constructed 231 ground-mounted 10 kW solar modules on member farms. The electricity generated is then sold under the microFIT program to the Ontario Power Authority. Due to delays from the grid constraints on many members' microFIT projects, the co-op is in the process of constructing 10 concentrated 500 kw "Solar Gardens" (away from the grid-constrained areas) under the FIT program. Six of these for a total capacity of 3 MW had been constructed as of December 2014. These co-operatives contract installation to private companies, but the panels are all member-owned. The co-operative model thus allows farmers to receive a higher return than the more common lease payments from larger private developers (AGRIS Solar Co-operative 2012, 2014).

Finally, for some projects, co-operatives are initiators rather than owners. These have, however, played important roles in developing projects, with

TABLE 7.2

Selected electricity generation co-operatives by province and structure, 2013

Structure	Province	Name	Type	Installed capacity	Progress
100% co-operative	ON	AGRIS Solar	Solar	5.3 MW solar projects	Constructed 6/10 500 kw FIT projects and 231 microFIT as of November 2014.
	ON	SolarShare	Solar	5.5 MW solar installed	33 projects contracted and built under FIT programs 2010–15 (130–600 kw each)
	QC	Co-opérative Forestiére de la Matapédia (worker co-op)	Biomass	0.50 MW and 0.80 MW	Completed 2009. $2.16 million biomass facility from wood waste to heat local hospital.
	AB	Irrigation Canal Power Co-op	Hydro	3 hydro plants (38.8 MW total)	Co-operative of three Alberta irrigation districts (Raymond, St. Mary, and Taber).
Community partnership	ON	WindShare Co-operative	Wind	0.75 MW (0.6 MW actual installed capacity) wind turbine	Project completed in 2001. Partnership with municipal utility (Toronto Hydro).

▶

◄ **TABLE 7.2**

Structure	Province	Name	Type	Installed capacity	Progress
Co-op–private partnership	QC	Val-Éo	Wind	50 MW	Received power purchase agreement from Hydro-Québec December 2010. Completion projected for Spring 2016. A 106 MW second phase has been proposed. Partnership with Algonquin Power and Utilities.
Co-op initiated project, owned by private partner	BC	Peace Energy Co-operative	Wind	102 MW windfarm	Completed in 2009. Selling power to BC Hydro. Partnership with Aeolis and AltaGas.
	NB	Lamèque	Wind	45 MW project	25-year power purchase agreement with NB Power completed 2011. Partnered with Acciona.

Sources: FCPC 2014; SEA 2011; individual co-operative websites 2010, 2011.

many hundreds of volunteer hours put into community meetings and project development, sometimes with financial investment as well. In the case of both the Peace Energy Cooperative (PEC) and the Lamèque Renewable Energy Co-operative (LREC), the co-ops initiated projects but do not own a portion of the project assets. With PEC, the project owner of the 102 MW Bear Mountain Wind Park (in Dawson Creek, British Columbia) is AltaGas. With the LREC (in Lamèque, New Brunswick), it is the North American subsidiary of Spain's Acciona Energy that owns and developed the 45 MW wind farm. The co-operatives formed to bring new renewable developments to their respective communities, and more than ten years of planning and development were successful in that aim. PEC received a finder's fee for its work in developing the project and negotiated a small investment piece (less than 1 percent) of the revenue stream from the project for co-operative members that chose to invest. Even so, the past president of PEC argued that the co-op "pushed very hard when the contracts were awarded from construction and equipment hauling so we'd have local businesses participate so we could generate as much economic spinoff as possible" (Rison personal interview, October 14, 2009). PEC is now involved in solar and geothermal projects in Dawson Creek, as well as in sustainability education.

The range of ownership stake in projects has implications for the co-operative difference. The higher the ownership share, the more control and economic benefit. This supports the co-operative difference, in theory, with investment and purchasing practice. Significant ownership shares are accompanied by higher financial returns and greater spinoff benefits (both financial and nonfinancial) in terms of local procurement and employment. Investment opportunities for residents in new power projects provide access to a revenue stream that is an improvement upon ownership that is highly concentrated in a few (nonlocal) companies. In one sense, this is democratization of ownership. However, many of the benefits of co-operative ownership also have to do with control beyond just investment. They have to do with democratic governance in making siting choices, as well as improved service provision.

Participation and Power
The community-based generation projects described above arose from the hard work of volunteers who were looking to create local economic development and to shift to new electricity and fuel sources. All of these projects involve ownership structures in which the local membership is engaged in decisions about the organization (often one member, one vote) and any profits are recirculated back to the community. This makes them different from a typical shareholder-owned energy-sector project. In cases like Wind-Share Co-operative, members had yet to make a financial return in 2012, more than eleven years after the turbine was built. The value in that particular project for many members was simply to build it in order to show that wind power in Canada was feasible and that local people could initiate and own renewable energy projects.

For several co-operative actors, the electricity co-op contribution went well beyond investment to direct control over projects:

The whole customer care concept is different; most utility companies today try to promote the customer as the biggest asset, and they love you, you're their biggest asset. In reality, the shareholder holds the power. They're there to make money for the shareholder. Don't tell me I'm the most important person to Fortis; I'm not. When push comes to shove, you'll find out who's the most important. (Nagel personal interview, November 27, 2009)

Members [in co-operatives] are the owners, the shareholders, and the customers. They're everything. (Bourne personal interview, December 1, 2009)

If we can get a big project here, what other value can we get out of it for the community? ... If it had been a strictly private enterprise, these questions would have never entered the picture. These are some of the values we can bring as a co-operative because we're not driven just by the co-operative, we're driven by the value we can bring to the community. Conventional corporations are mandated to be profitable for the shareholders – the shareholders who hold the most shares make all the decisions. In a co-operative, everyone who holds even one share gets an equal say; it's just more inclusive, and what's so wonderful about the co-operative model [is] it allows for the easy incorporation of social values – it's much more difficult with private corporations. (Rison personal interview, October 14, 2009)

Lessons about the value of project control from the experiences of distribution co-ops in Alberta are germane here. They demonstrate the value of actually controlling the co-operative assets rather than contracting out, and that once the facilities are built, the pressure to sell to private competitors increases (Chapter 6). Self-operating REAs are better placed to assess the real costs (and values) within the system when pressure to sell arises. The self-operating REAs had to fight for this control through costly legal battles and arbitration:

We had it in our mind we were not going back on a contract with TransAlta where they were operating our system. They overbilled us, they misbilled us, we couldn't complain about the service from the linemen, but it was what the cost was, and what we were losing was control. You don't like someone else making decisions on your destiny, and that's how it felt – we were just rubberstamping everything, and our members were paying for that. We got involved and said we would accept no less than being able to operate our own system and we want an expanded membership definition. (Bourne personal interview, December 1, 2009)

Self-operating REAs can bring cost savings with control. CAREA, for example, is part owner of Prairie Power (a wholesale purchasing company) along with other co-operatives, and through member ownership saved an estimated $3 million in 2010 for its eight thousand members when compared with investor-owned utility rates (CAREA 2010). Co-operative distributors are not incentivized by profitability to increase the amount of power sold. Unlike other power retailers, the co-operative needs to provide for the power needs of its customers, and not expand. Finally, the strength and expertise of the self-operating REAs allow them to support and advise co-op boards under pressure to sell assets (Bourne personal interview, December 1, 2009).

The close links, then, between the management, control, participation, and investment in projects are key to the full range of co-operative benefits in this complex sector. For the Alberta Federation of Rural Electrification Associations president, memberships in co-operatives over the long term create important community links that help scale up their power: "If you have control of your own destiny, like you do with an REA, the money stays within the community because it is part of it, and those same people are on the county boards, the school boards, so once it is sold to the power company, you lose all that" (personal interview, November 27, 2009). More than this, though, self-operating REA rates are lower than that of investor-owned utilities (IOUs) for rural customers (Bourne personal interview, December 1, 2009).

Education: Co-operatives and Renewable Electricity

A further manifestation of the co-operative difference is the contribution new electricity co-operatives make to community education. Electricity co-operatives serve to educate the public, both on the co-operative difference and on renewable electricity, in two ways. First, because of the level and nature of member involvement, volunteerism, and nonprofit orientation, the co-operative form facilitates member education and empowerment. The educational contribution of the Toronto Renewable Energy Co-operative's (TREC's) WindShare project is significant: "Members [of

WindShare] learnt about green energy strategies (renewable energy generation and conservation), about co-operative development, management and operations, as well as sustainability practices ... WindShare has shared its expertise, skills and knowledge with other green energy co-operatives, thus expanding WindShare's community of practice and spanning new communities of practice" (Duguid 2007, 288). In addition, "co-operative structures are appropriate for advancing green energy and energy literacy. Importantly, green energy co-operatives are a significant addition to the Ontario energy industry because they are organized at the community level, involve situated learning through participation, capture peoples' interest because they have a direct stake in the enterprise, and support green energy strategies" (Duguid 2007, 289–90).

TREC played a pioneering role in co-operative renewable-electricity generation by educating others in Ontario and the rest of Canada about how to (and not to) develop a project. TREC was the first nonprofit renewable energy co-op in Canada. The co-operative has now spun off other co-operatives (WindShare, SolarShare, LakeWind) that are focused on generating power from renewables, but TREC itself remains a nonprofit co-operative dedicated to promoting and educating about community power in Ontario. TREC and WindShare were also both involved in founding the Ontario Sustainable Energy Association (see Chapter 8) and have as part of their mandates a goal of public education and raising awareness of possibilities for new renewables in Canada. Many other co-ops in Ontario were spurred by the TREC/Wind-Share example, where members unfamiliar with co-operatives became interested in the co-op model for community ownership and democratic project governance (personal interview, July 20 and July 23, 2009).

Co-operatives make an educational contribution beyond their membership. A number of co-ops explicitly place public and community education front and centre in their mandates and run, for example, school programs and site tours, and provide free or low-fee consulting services in their communities (Procter personal interview, July 20, 2009). When a project is able to balance ownership and financial returns with these public education mandates, the co-operative difference in the power sector comes into sharp relief. Whether successful in building projects or not, these initiatives have contributed to developing more informed and aware constituencies. These understandings are important for informed renewables policies going forward (Barry 2012; Etcheverry 2013; Nishimura 2012).

This educational focus is an aspect of most electricity co-operatives – that is, distribution and generation co-operatives – as well as being the singular focus of new networking electricity co-operatives. For example, the focus of the Vancouver Renewable Energy Co-operative, the Toronto Renewable Energy Co-operative, and the Sustainable Energy Resource Group is on public education: raising the profile of renewables, community engagement, and project development. The need identified by these nonprofit co-operatives is not primarily economic (as it is with many consumer co-operatives) but social and, in particular, environmental. According to one renewable co-op director, the co-operative emerged because of a desire to do something concrete about changing energy use and supply. The director explains how "two things came together [in Ontario]. Obviously, one was environmental awareness of energy habits, combined with an opening of the energy markets for the first time so other people can actually produce power. Up until that time it was all public, so you couldn't. [It wasn't] so much that nobody's doing this ... it was about doing it differently" (personal interview, July 23, 2009). Thus, the co-operative is not formed to serve only member needs; it works more like a nonprofit community association toward broader changes in electricity policy and public practice.

Community projects can be used as demonstration projects – as educative tools to engage broader audiences – and the work of nonprofit renewable energy co-operatives can be important on this front. Co-operative projects often intentionally play a symbolic role in shaping the public perception of the possible. This symbolic value to prove new models are possible is often cited by participants and initiators of these projects. The former head of PEC noted:

We've had strong support from members of the co-op and other community members who are not members. People feel like they understand wind better, they get a thrill from the towers, and we have class trips for school kids to go up to see them. All of this helps build the idea of renewable energy. On all of those counts, community-based wind really did deliver. (Rison personal interview, October 14, 2009)

In the WindShare project, volunteer members as well as Toronto Hydro employees are involved in servicing the turbine and leading site visits for school and community tours. Two hundred thousand Torontonians drive past the turbine every weekday on their commute, and twenty thousand people visit the site each year on educational tours (Ferrari personal interview, July 23, 2009).

The educational and public focus of these new renewable co-operatives, interestingly, runs contrary to the literature on established co-operatives, which has pointed out the lack of community involvement and investment (Côté 2000). Several factors may contribute to the educational and outreach focus of new electricity co-operatives. First, electricity co-ops outside Alberta are a relatively recent phenomenon. So, in order to develop a membership base, a range of discussions with communities and regulators needs to take place. Second, many of the participants in these co-operatives are new to the co-op sector, and many are new to electricity as well. As a result, participation in the co-operative, through membership meetings, volunteers' work, and attendance at conferences lead to significant member learning. Third, the motivation for many new co-operatives went beyond financial returns to changing public understanding of renewables, and to shifting policy – as the interviewees recognized. Unlike farm-based consumer co-ops or financial co-ops, public understanding and public opinion is front and centre for many renewable electricity co-ops because the aim is benefits not only for the membership but also for society more broadly.

Pitfalls: Grids, Financing, and Stalled Projects

Renewable electricity co-operatives clearly vary a great deal, in some cases using the co-operative model to garner community support but with little or no control of the actual project. This is, in part, because roadblocks arise when co-operatives try to initiate projects. The project-development phase is full of potholes, and many co-operatives never make it to acquiring a power purchase agreement and getting grid-connected. From PEC in British Columbia to Pukwis in Ontario to LREC in New Brunswick, co-operatives run up against challenges with unclear co-operative legislation, financing, grid access, and competition with private corporations for sites. In Ontario, government commitments to the nuclear industry form a significant barrier to community grid access (FCPC 2013, 2015; Green Energy Act Alliance 2011; Stokes 2013b).

Frustration for community groups and co-operatives has arisen with the new development opportunities through liberalized power sectors as well as with competitive pressure from large corporate actors. According to the Federation of Community Power Co-ops, a particular barrier in Ontario has been

> gaming by the private sector whereby 100 MW projects originally intended for the RFP program were broken up into 10×10 MW projects in order to conform to the RESOP program rules. The frustration and limitations of

the program felt by community power proponents and other community-based organizations led to the formation of the Green Energy Act Alliance and the call for a comprehensive and community-based approach to renewable energy development. (FCPC 2013, 7)

What in theory is a marketplace open to small generators is in practice far more restrictive. For example, communities in Ontario and New Brunswick actively pursued projects for almost fifteen years before projects got built, and never in the way these communities had originally planned. Generation co-operatives that emerge from this milieu are often partnerships with private energy corporations, with significantly reduced community content and control.

Grid and Site Access Challenges
Co-operative generation projects across Canada have stalled. Although the uneven distribution of political and economic power in the electricity sector is the key driver, in practice, specific challenges have arisen, namely site access, lack of power purchase agreement, and lack of grid connection. Since most co-operative members are new to developing electricity projects, they are already at a disadvantage. Beyond this steep industry learning curve, however, these three main points of tension lead to extremely long – in some cases more than ten years – development phases during which communities lose momentum and burn out, as happened with the Positive Power Co-operative in southern Ontario and with Pukwis. In some cases, however, when the co-operative has institutional support (like a municipality or First Nation), there is enough momentum to keep the project going through roadblocks, as with the TREC/WindShare development in Ontario.

In the wind industry, there is fierce competition for sites where the wind blows regularly and with force. Those able to move the fastest and with the best connections to wind resource data and to policy makers secure the best sites. For co-operatives, this presents a real challenge. WindShare's former director, Evan Ferrari, points out that

capacity is always a problem with community stuff. Especially on the co-op model – one member, one vote. Democracy is difficult and it takes time. You have a sole shareholder in a company with a big bag of dough. They can move quickly, making decisions and moving on, working through committees. From our perspective, we want the oversight, want members to be supportive all along. If our board moves too quickly without the support of our members, we'll get kicked in the head. (Ferrari personal interview, July 23, 2009)

The result is that co-operatives incorporate but can't secure the more lucrative sites. Or they partner in order to get access to these sites and financing but dilute the community content of the project. When PEC formed in 2002 to promote renewable energy in the Peace region, it found that a Scottish developer had already secured the local site investigation permits. After a series of inquiries, the co-operative subsequently managed to secure the permit when it argued that BC regulators had not followed their own rules on granting investigative use permits (Rison personal interview, October 14, 2009). In Ontario, for Bala Energy Co-operative's Jeff Mole,

> the real challenge is getting access to the resource, getting applicant of record status. You've got the co-operative registered and incorporated, you've now got access to the resource, can do feasibility and create designs, and once you have a technically feasible design, you can create a business plan, and once you have that, you can go to the OPA [Ontario Power Authority] and get an offer to purchase the power. (Mole phone interview, July 20, 2010)

Co-operatives then have to spend a great deal of up-front time on legal fees, on articles of incorporation, and on articulating the co-operative difference before they can even get to project development. It is in this sense that the co-operative difference is actually a liability financially, particularly in provinces unaccustomed to co-ops in the mainstream (like British Columbia). Steve Rison, former president of PEC, says:

> We were reinventing the wheel here ... It seems that most co-ops in BC are either consumer or producer co-ops. They don't seem to do the same thing that we're doing. It seemed like the law firm was telling us like we had to be treated like a private company selling investment shares, that there is no difference between a co-operative and a private company selling investment shares and we had to jump through the same hoops ... We will have spent $20,000 just in legal fees. (Rison personal interview, October 14, 2009)

In Ontario, a former co-operative developer argued that "you go into a small business enterprise centre and the co-operative option is not even listed. If you're lucky you've the option of a not-for-profit. More commonly it is 'Do I want to start a limited partnership or a sole proprietor?' A lot [of the problem] is a fundamental lack of knowledge of the model" (personal interview, July 20, 2009).

Another significant challenge for these projects is obtaining permission to connect to the electrical grid so that the power from projects can be sold back to the provincial power authority or (in Alberta) power retailers. For one developer,

> the problem remains that a lot of community projects are not going forward because there are grid access issues. This is the case for First Nations projects; there are two where First Nation communities have not been able to get access to grids. That is one of the things that we weren't able to protect and would have liked to. (Personal communication, August 9, 2013)

These grid connection issues are driven in part by the need to upgrade, but also by the dominance of set-aside capacity for nuclear generation. For example, connection issues affected TREC's second wind development project, LakeWind. LakeWind was a 20 MW – ten 2 MW turbines – project near Kincardine, Ontario, started in 2005 as a partnership between two coops, LakeWind Power Co-operative and Countryside Energy Co-operative. The project has stalled despite having the wind data, completed feasibility studies, and secured land, because the Ontario Power Authority reserved a section of the grid for power coming from the Bruce nuclear plants. This area, called the Orange Zone, happened to be in some of the windiest territory in the province. This challenge highlights the issue of connectivity and transmission capacity (real or perceived) for community groups wanting to sell to the grid, as well as of the power of incumbent industry actors in grid allocation. Says WindShare's former director Evan Ferrari:

> The emails in our network are flying, saying, "We need to be ready in the event that the Orange Zone goes down." My initial response to that is, "I'm going to have a lot more grey hair by the time that happens." I understand that we need to be ready; I'm just not convinced that we can hit the ground running, because the regulations haven't been finalized yet [in July 2009]. (Ferrari personal interview, July 23, 2009)

According to the Federation of Community Power Co-ops, "after investing almost $1 million community-raised funds in pre-development costs, [LakeWind] was unable to obtain a FIT contract owing to grid constraints, and remains dormant, three years after making its FIT application, and eight years after securing its site" (FCPC 2013, 9).

What emerges, then – out of a very real need for system administrators to control and distribute new generation within capacity limits – is a set of connection waitlists that projects sometimes sit on for years. These lists are prioritized differently across the country – in some cases first come, first served, and in others by whichever projects are "shovel ready." In these tests, as with the capacity assessments in the first place, the processes are not particularly transparent, and co-operative actors are at an informational disadvantage. In Nova Scotia, the private power company (with its own projects) is also providing the information about where the capacity exists in the system. In Alberta, according to one co-op developer, the rush for wind meant that REAs and municipal groups were looking at "wind and forecasting ... but still, at end of the day, when you could connect to the grid in Alberta when they finally lifted the caps, it was flooded by the guys with all the money and the little guy got bumped out" (phone interview, April 13, 2010).

Financing Challenges

Co-operatives are at a disadvantage in meeting competitive calls for new renewables at the lowest cost, the electricity procurement mechanism most popular across the provinces. This is because groups are rarely able to raise the capital for large wind farms, and that is where the lowest kWh prices are (because of economies of scale). This does not, of course, mean that 100 percent community-owned projects are impossible, even for relatively large projects, just that private actors, particularly large ones, have key advantages in the marketplace, community FIT adder or no.

With the deepest pockets, they can build the largest and most lucrative projects, which in turn help with financing. Furthermore, prior experience with project development, as well as cash equity and capital assets, all play a role in structuring loan terms in such a way that disadvantages community projects. According to several community renewable energy consultants and developers, the result is that private energy companies are able to see projects to fruition where smaller, community-based projects run out of time, money, volunteers, and energy in dealing with the electricity sector:

> I don't think it is realistic that communities will build anything significant in this neck of the woods [Nova Scotia]. I don't think they can raise the funds.

There's no bank that will loan them the money, the government hasn't set up a fund for this, they've talked about it but haven't done anything yet – that might be a different story. They'll have a limit of $4 to $6 million, which you might get two turbines out of. Nobody knows who will be eligible, what kind of rate they'll charge. The real issue is not the cost of the equipment anymore; it is the cost to borrow money. If you get 75 percent of your capital costs at 8 percent, maybe it doesn't go; at 7 percent it barely goes, and ... all it takes is a 1 percent shift in the cost of money to turn these projects down. A community group with no track record will pay the highest interest rate, not the lowest one. That's what is happening to us; that's why we're looking for a funding partner who has a track record and can get money at a lower rate. They have access to money that is a full 2.5 percent less than what we pay. When NS Power goes to borrow money, they get it at 4.5 percent, when that goes up against a private producer, they'll win every time. (Zwicker personal interview, May 20, 2010)

Having the CP Fund [in Ontario] help support soft costs that's really hard for communities to come up with – like, $300,000 to do a feasibility study and a anemometer tower is great, but what about the capital once they get past that stage and they've got a project and have an agreement to connect to the grid? Let's say it's a $25 or $30 million project, let's say the debt-equity ratio is 20 equity and 80 debt – where are they going to get the 6 million bucks? That's not easy. Whereas food co-ops, $25,000, $50,000 to start up, no problem. And then you grow your equity. This is the reality in all businesses, co-op or not. They're self-financing through retained earnings. It's hard to raise the capital if the ROI [return on investment] is, say, 3 to 5 percent. Especially if that's what inflation is. (Personal interview, July 23, 2009)

We just can't pay $3.3 million per MW [in rural Quebec]. How can we find that? I didn't agree to put in $100,000 only to have a ticket to the lottery [to bid into Hydro-Québec's power calls]. In a little village like we have here ... it is just not the best way to spend our money. I was negotiating with companies who had won, and then ones who hadn't. Here only the Royal Bank was willing, but they needed an agreement with Northland Power from us and a HQ contract ... Out of eight community projects in Quebec, just one won. At the financial level, they threw us out. (Gagnon personal interview, May 16, 2010)

The persistent challenges of financing have implications for whether community power remains a marginal piece of the broader electricity sector. Connected to the issue of raising sufficient capital is that of lengthening the member-user link. One of the key virtues of the co-operative form is that it connects service users to production, with the attendant governance and educative and financial benefits this entails. Because of the nature of the good and the grid, new generation co-operative members are essentially investing in a business that sells to the public at large. In return, they reap financial rewards and, in some provinces, help reduce reliance on coal or nuclear generation. However, the direct co-operative link where members either produce the good or use the good (or both) is severed. The wider the circle the co-operative expands into to find project funding, the more diffuse the benefits, the connection to the physical assets, the ability to attend meetings and participate, and so on.

So, even if a project secures a site and funding, and navigates the legal and contractual maze that is power-sector development, it may still face connection challenges. Sometimes this is because there is already too much generation in the region. Sometimes the transmission and distribution infrastructure is old and needs upgrading. The wait times in connection queues kill the volunteerism vital to the community sector, and projects often do not have the money to sustain themselves. Funding supports to community actors in Ontario are currently not timed to coincide with FIT application windows, nor are they available to projects that have navigated early project development and been granted a notice to proceed (at which time financing needs to be secured) (FCPC 2013).

Community Capacity Challenges
One promising project that ultimately collapsed is the Pukwis Community Wind Park. After nearly eight years of development, by 2011, Pukwis had managed to overcome some of the largest hurdles – contracts with the Ontario Power Authority and project financing – facing community projects in Canada. However, by 2012, the project was cancelled. This was because of a combination of connection delays, regulatory hurdles, and general lack of familiarity with and flexibility for community energy models. Taken together, these resulted in what one co-operative developer described as "death by a thousand cuts" (August 9, 2013). This illustrates the toll that path-breaking takes on projects, ultimately undermining the will and ability of partners to proceed. On the one hand, these groups are providing innovative examples of ownership structures and engaging local stakeholders in

electricity development; on the other hand, they face a daunting uphill battle against deeply entrenched sectoral norms.

Developed as a partnership between the Chippewas of Georgina Island First Nation and the Pukwis Energy Co-operative, the Pukwis Community Wind Park was to be the first joint community and Aboriginal project in Canada, with a 51 percent First Nation share in the project and a 49 percent co-operative share. What made this arrangement significant in comparison with other co-operative wind projects is that this project was much larger than what others were managing to develop without large private partners. Phase one involved ten turbines (20 MW) at a cost of $55 million, expanding to 54 MW in phase two. The first phase was awarded a FIT contract in 2010 and included both the First Nation and the community supplements; this meant a contracted rate of 15 cents per kWh instead of the regular FIT rate of 13.5 cents per kWh. The revenues were projected to be $159 million over twenty years, going back into the First Nation and the co-operative members in the Toronto area for community development and further renewable energy projects. The project was unique in other ways too. The wind park was set to be developed on the territory of the Chippewas of Georgina Island First Nation, an island community of eight square kilometres and approximately 660 residents. One of the complications that arose from this was a jurisdictional battle as to whether environmental assessments on First Nations land are subject to provincial and federal environmental assessments. Ultimately, the project proponents were able to resolve the lack of clarity on this matter, but the process was both time-consuming and frustrating.

Provincial electricity policies also led to project delays. In 2003, when discussions between the co-operative and the First Nation began, Ontario had yet to implement the RESOP. Once these renewable development incentives were put in place in 2006, the project partners saw an opportunity to develop something both economically beneficial and technically feasible (in terms of connection to the broader system). Unfortunately for them and for many other community actors, the RESOP did not provide the facilitative framework in practice that it did in theory, because of issues with connection access, nuclear industry influence, and lowest-bid procurement. The introduction of the Green Energy and Green Economy Act in 2009 and its accompanying feed-in tariff program addressed many of these issues. When the contract was finally awarded, the partners and most of the community members were keen to move forward to completion; indeed, in 2011, the community held a referendum on the project, which garnered support from roughly 82 percent of the population (personal communication, August 9, 2013).

One important challenge that emerged for Pukwis was the rigidity of the contract milestones. According to one project member, there was no leniency or leeway in the meeting of target deadlines; this posed a challenge when the community groups hit unforeseen roadblocks that, because of either size or capacity, they were unable to overcome in a timely manner. For instance, by October 2011, the project needed to lodge a $1 million deposit for connection infrastructure from the turbines to the nearest transmission infrastructure. However, in May 2011, the Canadian government agreed to settle the Coldwater-Narrows land claim submitted by the Chippewas Tri-Council – which includes the Chippewas of Rama First Nation, the Chippewas of Georgina Island, and the Beausoleil First Nation. The settlement, which included the Chippewas of Nawash Unceded First Nation, was for a total of $307 million and required each community to go through a community-approval process that extended until July 2012. Given the small size of the First Nation, as well as the heavy responsibilities this placed on the chief and council, the timing of the Pukwis contract milestones was inopportune. According to one co-op developer:

> The land claim award came and it kind of turned things upside down on the island, and with chief and council in a critical time in the lead-up to these milestones that had to be met. It just became overwhelming ... It was such a long struggle dealing with this huge influx of capital in such a small community of people, each with their own idea of how the money should be spent and how much they should get ... The entire summer was consumed with those issues and community meetings and some of them quite heated, and the issue at hand for community members became far more important than the development of a windfarm. (Personal communication, August 2013)

The injection of funds the claims settlement provided to the First Nation was certainly beneficial, yet it is easy to overlook the fact that communities have limited time and capacity. Only so many large projects can be juggled. This may point to a limitation to the expectation that community-led and owned projects will play a large role in any given electricity system. Or it may suggest that system planners need to incorporate more flexibility for small actors. In the summer of 2011, the project partners asked the ministry, Hydro One Networks, and the Ontario Power Authority for more time to meet the interconnection payment and milestone given the circumstances. It was not enough to save the project. Despite a good faith payment of the $1 million deposit by a partner co-operative, the First Nation decided the timing was not right to proceed. As one project member put it, "We needed six months to a year, and they gave us three weeks" (personal communication, August 9, 2013).

Compounding the timing challenge was the professional advice project members were given throughout the process that co-operative structures are "not the way energy projects are done in Canada." This opinion was expressed by federal and provincial actors, who suggested that more conventional business models would be more appropriate vehicles for project development. This went on late into the last days of the Pukwis project. Ultimately, it was this litany of challenges that killed the project: the long delays at the early stages, the many regulatory and infrastructural hurdles, the need to deal with claim approvals, and the deep skepticism about co-operatives. Even if all the other hurdles are cleared, the issue of milestones, timing, will, and capacity persist.

Partnerships: Both Promise and Pitfall?

The challenges confronting co-operative generation projects have resulted, across Canadian provinces, in numerous co-op projects that are either partially or wholly owned by conventional private-sector actors. There is a range of project ownership models, from ones wholly owned by co-operative and community groups, to joint partnerships with municipalities or First Nations, to minority stakes in projects with large shareholder-owned developers. Partnerships with organizations that have experience and funding access allow for the development of larger, more lucrative projects, and often a more streamlined process, since private partners tend to have development experience. Attempts to form electricity generation co-operatives thus provoke a dilemma for members: partnerships help smooth the project-development process, but they also (in most cases) dilute the community control and return. Many are left hoping that "angel" development companies interested in their public profile will develop the projects and allow for increasing levels of community investment over the life of the project (Loring 2007; N. Meyer 2007).

There is a trade-off between the efficiency of expert developers and economic democracy and control. There is a danger in minority partnership models of eroding any real difference that co-ops may bring to the power sector. Gordon Walker, commenting on the UK experience with co-operative electricity projects, argues that private-sector domination of community partnerships is ultimately damaging to sustainability in the power sector (Walker et al. 2007, 78). What emerges is a system where private developers "game" communities and the co-operative difference for financial gain. Says one former co-operative developer:

> Developers thought community projects were cute. I got the sense that they didn't need to drive us out of business because they never took us seriously in the first place. Several co-operatives tried to partner with multinationals in their area ... You know, let's use their resources, the community will get additional dollars, and it'll make it easier for them to develop because there will be community ownership, and when the co-operative went down this road, the multinationals said, "You're talking about social capital stuff and that's not real dollars. Unless you can hand us the community on a plate [it is not worth partnering], or we'll give you a kickback to shut you up," but there was never any sense of real partnership in those. There were a number of those "We could carve off a spot for you, we'll develop ninety-nine turbines and you can have one to community finance." (Personal interview, July 20, 2009)

Both AltaGas's Bear Mountain wind farm and Acciona's Lamèque wind farm are arrangements where the co-operatives played a key role in bringing the project to the community but ultimately own small (PEC) or no (LREC) shares in it. In both cases, local co-operative partners initially set out to develop their own generation project, but the roadblocks (financing in particular) proved too significant to overcome. Steve Rison, former president of PEC, notes:

> AltaGas makes most of the money. But we will get some money locally, and I think that's a key difference between, as I said, ours and the one in Chetwynd [Dokie], where the locals don't get a dime, [just] all of the negatives and none of the positives ... I negotiated with Aeolis [the other private partner] that if you get a developer fee, we get a piece of that in recognition of the value of the community support we brought to the project; we'd already done that work. They recognized the value because they were trying to develop another wind project in the very early stages off an island near Vancouver and there was a lot of resistance in the community. When they came to Dawson Creek, there wasn't any resistance because we'd already built the groundwork for support, had support from city council, had support

within the community. So it was a very different situation for them. (Rison personal interview, October 14, 2009)

This cautionary argument about partnership needs to be qualified, however, as not all partnerships are the same, nor are all partners. The need to partner varies depending on the range of challenges (funding, expertise, sites) the new co-operative faces. In cases like that of Val-Éo, local landowners already had the site secured and they were able to leverage their considerable assets (including years of wind data) to negotiate a majority partnership with a 75 percent stake. Significant community ownership also exists in projects where the partner is a municipality or First Nation. The local roots are retained, the revenues are recirculated back into public projects, and elected agencies have control over both project design and revenue disbursement. Most municipalities and First Nations are not expert electricity project developers, yet partnering with them helps balance the need for institutional support (outside a volunteer base) over the long term with a desire to preserve community control.

Partnerships with municipalities can help ensure that revenues and control stays local. Indeed, the pioneering work of the Toronto Renewable Energy Co-operative in developing WindShare was coming together as a 50-50 partnership with Toronto Hydro. The municipal utility brought sector experience, funding, and a range of other strengths to the project. These projects, municipal or otherwise, need not be small. In Ontario, several new co-operatives have followed suit, including the Sustainability Brant Community Energy Co-operative, which partnered with the municipality and was awarded a FIT contract to develop 350 kW of solar power in 2013. Looking outside Canada to Copenhagen, the Middelgrunden wind farm is a 50-50 partnership between the municipality and the Middelgrunden Wind Turbine Cooperative, with ten thousand members. When it was built in 2000, it was the largest offshore wind farm in the world, with a capacity of 40 MW, generating 4 percent of Copenhagen's power (Vikkelsø, Larsen, and Sørensen 2003).

The pressures to partner are most significant for new generation co-operatives. Electricity retail co-operatives do not face the same challenges, nor do those working on community education and at the volunteer project-development stages. The issues around control and power are important, though. Distribution co-operatives in Alberta (other than the seven self-operating ones) have, for the past three to four decades, had their assets managed by private power companies. The consequences have been problematic for co-operative financial strength and durability. The alternative partnership models outlined here suggest that for co-operative power generation to succeed on a broad scale, a range of other local and public actors need to become engaged.

Summary

Power generation in Canada is big business and power-sector restructuring, together with new renewable-generation incentives, encourages foreign and domestic investors to compete for sites, contracts, and grid access. Canadian electricity-generation co-operatives are spurred on by new policy developments but motivated by community actors keen to own a share of emerging private renewables sectors. Co-operatives bring significant benefits to projects, including local employment, project control, cost savings, and education. However, the challenges are significant. These projects are not taking place in unprofitable and underserved areas as early rural electrification did. As a result, real differences confront already-established distribution co-operatives and newly developing renewable generation co-operatives. Both are under pressure from other independent power producers (IPPs) and powerful incumbent industry actors – investor-owned utilities in Alberta and nuclear interests in Ontario. In generation, the pressure is to partner or, at the far end of the spectrum, for community and co-operative actors to facilitate but not own the project. In order to get generation projects built and connected to the grid with a contract to sell power, the community ownership and control of projects is often watered down. What is in principle an open market for IPPs to sell electricity in practice favours companies with prior experience, deep pockets, and industry connections.

8

Co-operative Networks and the Politics of Community Power

Electricity co-operatives, as with the co-operative sector more broadly, find strength and power in their networks and associations. These networks facilitate project learning between a range of provincial and sectoral actors and are helping new electricity co-operatives to scale up. That is, electricity co-operatives today are far from ad hoc one-off projects; they are part of a movement. Mobilization by electricity co-operatives has led to provincial community power policies in Ontario, Quebec, New Brunswick, and Nova Scotia. Mendell and Vaillancourt, writing about the co-operative sector in Quebec, have argued that co-operative associations can be, and indeed have been, involved in "co-construction of public policy," where social economy and co-operative actors play important roles in developing as well as responding to state policy (Laville, Levesque, and Mendell 2007; Vaillancourt 2008). These developments support the contention that a co-operative power movement is growing and is powerful enough to drive new policies.

This movement is accompanied by significant challenges, including but not limited to new policies in some provinces being far from effective in ensuring community control of projects. In addition, the breadth of the definition of "community power," together with the tendency toward project partnerships with private developers, weakens the ultimate contribution to economic democracy and empowered participatory governance that this movement can make. Thus, a crucial challenge going forward is defining "community" and setting policy so that the local and public participation aspects are strengthened and communitywashing does not become the norm. The degree of political mobilization and networks, as well as depth and strength of a corresponding political movement, contributes to countervailing power. This, along with participatory governance structures, sets the foundation for empowered participatory governance (Fung and Wright 2003; Wright 2010). It is important that these networked movements not only attempt to make changes but have the organizational strength and rhetorical force to be successful in their attempts. As a result, the motivations, network participation, and political role played by co-operatives matters.

In Ontario, the community power movement that emerged out of the Toronto Renewable Energy Co-operative project is strong, growing, and spilling over into other jurisdictions. Renewable generation co-operatives straddle membership in both co-operative tertiary organizations and private-sector networks like the Canadian Wind Energy Association (CanWEA) and the Canadian Renewable Energy Alliance (CanREA). In 2013, a new Federation of Community Power Co-operatives was formed in Ontario in order to advocate for the sector. By tracing the development of the strongest community power network – Ontario's – I illustrate how electricity co-operatives, in concert with the broader renewables sector, were central to policy change in that province.

Motivating Community Mobilization

Why have people formed co-operatives to develop new power generation in the past ten years? Why would communities *want* to mobilize and volunteer to build new generation when most households have access to power? Co-operators have three primary bottom-up drivers. First, people starting these co-operatives are motivated by a desire to change their provincial electricity mix toward more renewable fuel sources, such as wind and solar power. Today's generation co-operatives want to promote wind, shift from coal and nuclear, and prove that community ownership in electricity can be done. George Alkalay, a co-operative adviser and consultant, put the motivation for AGRIS Solar Co-operative (solar generation) members this way:

> One of the things that people were talking about very openly and publicly was, rather than having all these farm and European consortiums coming in making those profits and taking them out of Canada, why don't we the farmers own the electrical generation capacity and keep those profits back

in Ontario ... Is that transformative? Is it restorative? Is it reactionary? It is more than just a business model ... there's something about that democratic control that is really critical. (Alkalay personal interview, April 15, 2010)

A second major motivator for co-operative energy projects is community development. When public policy allows for private ownership of generation resources, some communities have opted to compete in order to control what they consider local and public resources. In the absence of public control, co-operatives form to provide an alternative. In Quebec's Gaspé Peninsula, renewable energy and electricity co-operatives have been emerging in poor rural areas where local co-operative development agencies (i.e., RDCs – regional development co-operatives) are facilitating them. Corporate wind developers have been building hundreds of turbines in rural areas as a result of the province's private wind-power calls; local people have responded in order to use, develop, and control their own energy resources and to stimulate local economies. There is pressure to move quickly, however, before large private companies secure remaining resources and lands. Martin Gagnon, the director general of co-op development in Bas-Saint-Laurent/Côte-Nord, described the process in 2010:

The co-ops are usually created when the economy is down and when people are exploited. We're in a region where we lost the fisheries twenty-five years ago, and we were losing agriculture here and passing through the worst crisis in the forest industry. Our region was and is the poorest region in Canada. The people are beginning to understand that their lands and the resources can be developed by themselves, not by any private group from anywhere in the world. That's the problem, the people are always thinking that it is God, the government, or foreign private (companies) that will come and develop our community. So, little by little, this wind-power development in Lower St. Lawrence is one of the lights, the flash that is going to open the door. The wind turbine is big, creates an impression, and to put that on the land of a guy who is dying because he doesn't have anything to live on is too much. The people are looking at that – and they receive only $500 per year – and wonder, If developers from Toronto or Edmonton can do this on my land, why can't I? So the people are [becoming aware] of that. We've done more than a hundred public assemblies with people to explain this kind of economy and to understand what is the [role] that they can play in [these opportunities]. So it begins to work ... After more than three hundred years of history in this region, we had only two multinationals. In the last six or seven years, we've had more than twenty-two private multi-nationals here to do business. I think it is going to wake people up. (Gagnon personal interview, May 16, 2010)

This is also the case with the Bala Energy Co-operative in Ontario, which was formed to develop resources the private sector was about to take over. One of the founders described the motivation this way:

> Water is a public resource, the land used in this case is public land, and the public is going to suffer the impacts [of development]. The public has to buy the energy at incentivized prices because of the feed-in tariff program. For those reasons, I thought the public had a lot of skin in this game and wasn't getting a lot [out of it]. So how can you still develop the energy resource but also provide a vehicle for developing it for the community? Municipal ownership was an option as well. I spoke with the municipality. The previous mayor gave me a lot of reasons why they couldn't and wouldn't do it ... so we set up a co-operative. (Mole phone interview, July 20, 2010)

For provincial policy makers, the reasons differ somewhat. Supporting co-operative and particularly *community* power has two important benefits: service provision and legitimation (Nova Scotia Department of Energy 2014; Nova Scotia Government 2011). In the period of rural electric co-operative development, the policy supports allowed provincial governments to address public concerns and to meet public needs without the appearance of state interference in the market. That is, co-operative development was ideologically compatible with free-market policies and could be used to fill unprofitable, albeit necessary, state functions. Using co-operatives in this way entails an appearance of democratizing decision making to meet local needs, without long-term direct public investment. This service-provision function of co-operative-state relations remains today in other sectors also, for example, housing, health care, and food.

More recently, community-based (including co-operative) power policies in Canada have been justified through the legitimizing role they play in other government goals. A public servant in Nova Scotia's Department of Energy (Nova Scotia Energy personal interview, May 19, 2010) pointed out that the province's community power policy was put in place in response to a lack of renewables support for other projects: "[Community power policy] is one of the vehicles for getting more community acceptance for renewable energy development projects ... Before we lost momentum in our renewable energy policy, we wanted to get something that was seen to maintain community acceptance ... it'll be an opportunity and not a guarantee." So although community power can help overcome opposition to the cost increases associated with wind and renewables, thorny issues of the definition of "community" persist. Walker et al. (2007, 78) caution that "perhaps the critical judgment here is the extent to which the 'shallow' use of the term community, to include essentially technical projects with minimal local collective involvement or benefit, is corrosive of deeper principles of socialized, locally-led and owned distributed generation."

Networks: Co-operatives and Community Power

As a result of this range of motivations, today's electricity co-operatives are networked with diverse groups: co-operative associations (e.g., Canadian Co-operative Association), new renewables associations (Canadian Wind Energy Association, Canadian Renewable Energy Association), and more recently community power associations (Ontario Sustainable Energy Association and Federation of Community Power Co-operatives). Many (but not all) are connected to provincial co-operative networks. Co-operative umbrella federations exist at the provincial level to facilitate, among other things, co-operation among co-operatives; they then federate up to the Co-operatives and Mutuals Canada and the Conseil canadien de la coopération et de la mutualité. For example, Battle River Rural Electrification Association is a member of the Alberta Federation of Rural Electrification Associations and the Alberta Community and Co-operative Association, which then federates up to Co-operatives and Mutuals Canada.

Co-operative associations, whether provincial, federal, or sectoral, affect the degree of penetration and resilience of organizations in the movement. They also contribute to resilience when sectoral threats emerge. The federations of rural electrification associations and gas co-operatives in Alberta have lobbied successfully for policy changes that allow them to continue operation and expand beyond serving only farmers. As well, larger REAs have helped smaller ones withstand pressure to sell their assets (Bourne personal interview, December 1, 2009) and are now lobbying the government of Alberta to change the rules so that REAs are not forced to sell to private investor-owned utilities (IOUs) rather than to other REAs when they demutualize. The Federation of Community Power Co-operatives is involved in lobbying the government to facilitate RRSP eligibility for energy co-operatives, to help overcome financing challenges (FCPC 2013).

Co-operative electricity associations and networks also serve to provide the framework for "co-operation among co-operatives," one of the seven international principles of the co-operative movement. Institutionalized networks do not erase the many challenges co-ops face, but they do ameliorate some of the effects of being a relatively small player in the power sector.

Outside Alberta, co-operative associations have not had much to say about electricity co-operatives until very recently (CCA 2011a). In Ontario, the Ontario Co-operative Association started to aid their development with feasibility studies and co-operative education (personal interview, July 20, 2009), but in the Canadian co-operative sector as a whole, the lack of new co-op development in this area until the late 1990s resulted in scant attention. As environmental sustainability issues have become more popular, the profile of electricity and energy co-operatives in the co-operative sector has started to change. Co-operatives and Mutuals Canada now includes a sustainability committee, and research has started to emerge on the co-operative contributions in the areas of biofuels, sustainable food, electricity, and heat cogeneration. The committee has been working with a range of university researchers and actors on a five-year (2010–15) "measuring the co-operative difference" national research project, culminating in an online resource hub for information on the sector at www.cooperativedifference.coop.

Community power networks have been the most active groups at this point. Some electricity co-operatives are also part of US and European, as well as domestic, co-operative and renewables networks. CAREA members, for example, attended REA conferences in the United States, and at one point CAREA owned a power wholesale purchaser, Prairie Power, as well as a western Montana generation and transmission co-operative (Bourne personal interview, December 1, 2009). These links have helped co-operatives under pressure to sell their assets; the power wholesalers help keep electricity prices lower for REA members, and the associations serve as useful lobbying mechanisms. Co-operatives in the power sector are networked too within a broader community power sector comprising First Nations, farmers, co-operatives, municipalities, and nonprofits. New provincial policies are targeting this broader sector, rather than co-operatives specifically, so it is this broader sector that has some degree of policy influence. This is particularly true in Nova Scotia, where Community Economic Development Investment Funds have garnered the lion's share of contracts.

Community power networks in Canada are also tied to the renewable electricity associations, such as the Ontario Sustainable Energy Association (OSEA) and the BC Sustainable Energy Association (BCSEA). OSEA, for example, together with other community and environmental groups – the Toronto Renewable Energy Co-operative (TREC), the David Suzuki Foundation, Environmental Defence, and the Pembina Institute – was part of the Green Energy Act Alliance in Ontario. The community networks have emerged in order to meet a multiple bottom-line framework of increasing renewable generation, facilitating community economic development, and increasing local participation in power projects, particularly in provinces where private power generation is facilitated. In fact, for one co-op director,

> it wasn't the co-op sector that was driving [the Green Energy Act] at all. It was the environmental community and the community power sector, which goes beyond co-ops ... There were a few individuals that represent co-ops, but for the most part, there aren't really co-ops other than TREC and perhaps five others, only one of which has actually built a project. This is much more about an environmental community coming together. (Personal interview, July 23, 2009)

Tracing Canadian Community Power Networks

The development of the community power sector in Ontario illustrates the power of policy networks and supportive institutions, as well as interjurisdictional policy learning and organizational practice. The province has become the hub and model for Canadian community and renewable power advocates across the country. Key initiatives include annual community power conferences, start-up funding for projects (via, for example, the Community Energy Partnerships Program), webinars on best practices, and the creation of a network of domestic and international legal and policy experts. Below, I identify the actors within growing community power networks, tracing their development from Denmark and Germany, through Ontario, Quebec, and Nova Scotia. The final section of this chapter raises important political challenges for these networks and the co-operatives within them based on the diversity of partner/ally goals in the community power sector.

TREC had part of its genesis in the Ontario Green Communities Initiative in 1994 and in the Energy Action Council of Toronto, insofar as that initiative helped bring together a group of environmentally conscious residents. Evan Ferrari, chair of WindShare Co-operative in 2009, managed the Guelph pilot of this program. He says that it was "instigated in 1992 or 1993 under Peterson Liberals, then the NDP came in and thought it was great, so they continued it. [It was] essentially a community project that looked at energy water and waste issues ... at energy audits. This sounds very matter-of-fact, but at the time it was groundbreaking" (Ferrari personal interview, July 23, 2009). TREC founders were inspired by the Danish experiences with community wind-power development and set about to develop a generation project of their own. Ontario's heavy reliance on coal and nuclear power sources spurred some members to action, as did the Harris government's restructuring of the Ontario electricity sector in the mid to late 1990s.

As with distribution co-operatives before, Canadians looked abroad – this time to Germany and Denmark – for electricity co-operative models. As mentioned, for the founding members of TREC, the Danish experience of developing wind power using co-operative organizations served as an inspiration and aspirational model. In Denmark, as in Ontario, there was a shift away from nuclear and a search for alternatives. The Danish case is often cited as an example of how electricity co-operatives help overcome the "not in my back yardism" – NIMBYism – commonly associated with power (and industrial) development (Gipe 2007, n.d.; Mitchell 2008; Toke, Breukers, and Wolsink 2008). By giving locals a stake in the profits and a say in the development, the co-operative form significantly reduced opposition to the look and noise of turbines.

Significant actors within Canadian community power networks have met with Danish (Preben Maegaard at the Nordic Folkecenter) and German (Henning Holst and Hermann Scheer) experts, and some have trained at and visited the Danish Nordic Folkecenter, as well as Schleswig-Holstein in Germany, including Patrick Côté of Val-Éo; Janice Ashworth of Ottawa Renewable Energy Co-op (formerly Ecology Action Centre); Kristopher Stevens of Centre of a Circle Consultants (formerly OSEA); and Paul Gipe of Wind-Works. One result of this training was an understanding of the many ways, whether through formal co-operative incorporation or not, that communities can secure significant negotiating and development power in new generation. According to one Ontario sustainability advocate,

[Val-Éo's Côté] created a really cool hybrid where the co-op structure isn't so much about owning the project, it's about forming a land monopoly. Bringing everyone together and saying, "We all agree that we will work together in creating a social contract; this contract then binds us ... when we go to the stage of share offering, or partner with other corporations to form an LLP." So the co-operative is the 51 percent shareholder of the LLP with the private developer. Then the local citizens get to invest in the LLP. (Personal interview, July 24, 2009)

Another result of these policy networks was the Green Energy Act in Ontario. In 2008, OSEA hosted the World Wind Energy Association conference in Kingston, Ontario, where community power was featured prominently. The new Ontario energy minister, George Smitherman, attended, as did David Suzuki, Herman Scheer, and a range of actors from Ontario's community power sector. Members of the Green Energy Act Alliance organized for Smitherman and key staff to go to Germany, Denmark, and Spain to investigate the potential for new renewables in Ontario (personal interview, July 23, 2009). What emerged from that interaction was a move away from the Ontario Power Authority's conservative twenty-year plans, and a clear rejection of coal-based generation (but not nuclear).

In 2013, a group of Ontario co-operatives formed the Federation of Community Power Co-operatives. It, together with the Community Power Fund, Friends of Wind, OSEA, Rural Ontario Community Power Producers Association, and Ontario Co-op Association, circulated a Community Power White Paper in 2013. In it, they call for three key changes to the Ontario sector: (1) a 1,000 MW procurement target by 2018 for co-operatively led projects, (2) a rolling COMFIT with priority access, and (3) the establishment of an Ontario community energy foundation to provide both early- and late-project-stage development support (FCPC 2013). The Federation also conducts annual member surveys of assets held, employees, co-ops in development and generating power, as well as the motivations and challenges for co-ops in the province. This resource is invaluable given the challenges of data collection on electricity co-ops outlined throughout this book.

From TREC to OSEA, NSEA, and BCSEA

The Toronto Renewable Energy Co-operative's WindShare Co-operative and SolarShare Co-operative projects are significant developments in the Canadian community power sector. From them, a degree of bootstrapping has taken place, in which referring to and building on this project model has led to many new co-operative projects (though not necessarily successful ones). TREC and WindShare members have been instrumental in educating and providing an operating model for community power in Canada. For example, a 2011 initiative of the co-operative, the Community Power

Investment Platform, provides a library of legal and financial templates for communities seeking to form power co-ops. Their hope in developing this tool was to prevent these new co-ops from costly and lengthy start-up processes. TREC members have played roles in developing the now defunct Ontario Community Power Fund, the Federation of Community Power Co-ops, OSEA, and, out of that, the recent Green Energy Act. Deb Doncaster, for example, was the executive director of the Community Power Fund and a founding member of TREC. According to one interviewee,

> the campaign [Green Energy Act Alliance] was the brainchild of Deb Doncaster, who was one of the first employees of TREC and then was the executive director of OSEA ... [She then] formed the CP Fund, which [was] the funding body for community energy projects. When the RESOP was cancelled, there was recognition from a number of groups that said, "We need to get to 100 percent renewable; we need an act to facilitate that." A number of people got together: the CP Fund, OSEA, Environmental Defence, the WWF, and then they brought in others like the Ontario Farmers Association and the First Nations Energy Alliance. (Personal interview, July 23, 2009)[1]

Indeed, many of the core players of the Ontario community energy sector have ties back to TREC and the WindShare project. Joyce MacLean, formerly with Toronto Hydro (the WindShare partner) was also on the board of TREC. Brent Kopperson, executive director of the Windfall Ecology Centre, is a founding director of OSEA. Paul Gipe is a community and wind-power expert based in California and an OSEA policy adviser. These networks have strengthened supports, incentives, and awareness of community and co-operative power across the country. Ontario now has one of the most supportive policy environments for these projects in North America. There are also links between OSEA and larger renewables associations in the country. According to one Ontario interviewee:

> CanWEA or CanSEA [Canadian Solar Energy Association] ... they're not [really] members, they may be members of the alliance, but not with any say. We consult them to say how we're going to comment ... Ultimately, we want the same goals, except that we want a whole lot of community power, whereas I don't think they really care – it's all about getting power into the ground and making money. Hey, that's fine, there's a lot of money to be made from power. We understand that if you don't want the community backlash, you need the

community to own it. We take a lot of our direction from community power groups. (Personal interview, July 24, 2009)

This was in some ways problematic, since the OSEA position, although it helped facilitate community power, has failed to secure significant space in a crowded private power market:

I think OSEA made a strategic decision at one point that they couldn't just limit it to co-ops. They had to be open ... the standard offer contract had to be open to any kind of business structure. You know what happened: private investors speculated and they didn't necessarily sign standard offer contracts, they tied up all the interconnection. They were just speculating, said, "Hey, we've got the interconnection point at this place, we're going to sell that to the highest bidder." (Personal interview, July 23, 2009)

Other provinces are now borrowing from the TREC/WindShare and OSEA expertise. In British Columbia, the Peace Energy Co-operative consulted with TREC in its development. And its private partner was connected to the German renewables sector and to TREC:

Jeurgen Peutter, the founder of [Aeolis, a project partner] had hired T.J. Shur, who had worked for TREC. She had just finished working for them, got her master's degree, and just gone to work for Jeurgen and Aeolis. She was keen on the co-operative approach, knew what it was like to work with a co-op – she was keen. They were a small start-up with three employees at the previous companies. (Rison personal interview, October 14, 2009)

In Nova Scotia, a Nova Scotia Sustainable Energy Association (NovSEA) alliance was heavily involved in pushing for the 2010 Community Energy Policy. David Wheeler and Michelle Adams from Dalhousie University developed convening materials for community input. According to Janice Ashworth, former member of the Ecology Action Centre and current operations manager for the Ottawa Renewable Energy Co-operative, the NovSEA alliance included a range of actors (including her): "Tim Weis at Pembina; Sierra Club Atlantic; Mike Layton at Environmental Defence; the United Steelworkers; First Nations groups; Nova Scotia woodlot operators; Scotian Windfields, [a CEDIF] company; and Seacor (Ashworth personal interview, May 21, 2010). Both Tim Weis and Mike Layton (and their organizations) are also part of the Ontario Green Energy Act Alliance.

These provincial sustainable energy associations (BCSEA, OSEA, NovSEA) provide education, advocacy, and networking between community and co-operative electricity advocates. Often run by volunteers and members, and through donations, they play a key role in popularizing and helping communities mobilize. In 2009, OSEA was talking with BCSEA and looking toward "massive rollouts" of community energy policy across the country (personal interview, July 24, 2009). Each year, OSEA sponsors the Community Power Conference, which draws hundreds of community leaders, utility and government representatives, farmers, and sustainable energy entrepreneurs. The networking activities in the sector are potentially powerful, creating new constituencies that can shape energy policy across Canadian provinces.

Politics: Communitywashing Power

As illustrated in Chapter 5, the development of renewables has become tied to private-sector expansion across Canada, so challenges have emerged with the expansion of opportunities for new market entrants. The question remains, will co-operatives continue to play a legitimating role in this, and what kind of role will it be? Two important political challenges have emerged within community and co-operative power networks: communitywash and NIMBYism. Community development and environmental groups have a great deal of enthusiasm about community ownership and community power (CCA 2011a; Community Power Fund 2010; Gipe 2010; Girvitz and Lipp 2005). Without a clear idea and definition, however, "community" can mean just about anything and, as we have seen, can sometimes amount to very little actual project control. The definition, then, is political and important.

The porous boundaries that exist between the community power sector and private power developers are problematic. Given the trend in provincial power sectors toward more private renewables, community power advocates are looking for a way for some Canadians to get at least a piece of that. Although some are still aiming for wholly owned community power, others are more conservative about what is politically possible, while doing their best to lobby and educate about the community alternative. One recent issue that has emerged is that under FIT 2.0 and 3.0 rules, contracts will go into default if the community economic interest drops below 50 percent. The impact of this is that long-term lenders are then far less likely to provide financing (FCPC 2014). However, there is a challenging balance that needs to be negotiated between meaningful community interest in the long term and diluting the rules to conform better within systemic constraints.

Within Ontario, many interviewees raised the challenge of the distribution of benefits between new renewables actors. Evan Ferrari, former president of WindShare Co-operative, describes his experience of being at the Canadian Wind Energy Association conference and interacting with the wind industry for the first time:

> I was at the CanWEA conference, and it surprised me because I hadn't been at an industry event before, hadn't been involved with the industry side of it to any extent. I was astounded by how Bay Street it was. This is not the tofu-sucking, sandalista-wearing folks that normally do this stuff. That blew me away. You're walking around looking at people's name tags ... I was embarrassed by the number of people who came up to me and said, "You don't know what impact you've had on this industry" ... The downside [for WindShare] is that we haven't been able to turn that into money. The rest of the industry has had a huge benefit. Our members, in spite of the huge obstacles that we've had, I have to give them really bad financial news. (Ferrari personal interview, July 23, 2009)

Another challenge is that of private-sector actors gaming community power financing supports and policies to support their own bottom line. This includes the concern, for example, of a "co-operative of five guys" (which is possible and emerged in Quebec), or a co-operative of corporations, or co-operatives and community power organizations from urban centres that are developing rural lands without local community buy-in. Initially, some renewable energy developers were interested in the co-op model because of differential rules for raising funds:

> Some of the Bay Street financiers thought ... "Hey, no $300,000 prospectus, let's take another look at this co-op thing," then they would find out that the CCA allowance wouldn't work for capital cost flowthroughs and you won't get a 15 percent return on investment ... so they said, "We're out." So you had all these people trying to turn the co-op into something it wasn't. There's not a shared idea ... the word "movement" makes people really uncomfortable. (Personal interview, July 20, 2009)

Another way communitywashing can occur is that

> you get a differential definition of "community." The more you get [support-
> ive policies] developing, the more you get groups trying to game the sys-
> tem. One thing brought up, which has not happened with wind so far that
> I know of, is a ... private corporation often from somewhere else lending
> money to local people with no interest and then reinvest[ing] in the project
> so that they get the adders and access to grant funding. That happened with
> a big agricultural grant with the feds. The co-op movement was up in arms
> about this. Some agriculture company just lent the money to the farmers
> with no interest, and they used that to buy part of the company. [So there
> are] suspicions about community power. (Personal interview, July 20, 2009)

Community power is, therefore, defined in different ways by different
agencies. Careful attention needs to be paid to where the lines of community are delimited and how much of an ownership stake is required. Negotiations between differing interests within and between communities require
planning and support. This recognizes that not all communities and groups
within them are equally well resourced or equally powerful. OSEA tried to
set minimum levels of community involvement as a recommendation in the
Green Energy Act. This discussion about what constituted community
power had OSEA members "tearing their hair out" in 2009. According to an
OSEA interviewee:

> One of our points was, before we agreed to have a minimum of 10 per-
> cent equity held by the local community, before we got to that point, it
> was "Well, let's say that TREC (just for example) finds an awesome wind
> regime somewhere." We go in, say to the local citizen, "Hey, you have the
> opportunity to invest in this project," and they say, "Screw you, we don't
> want that in our backyard." The discussion was, right now you get an adder
> for community power; just imagine that this external group goes in and
> says the community doesn't want to buy any shares, but we still have the
> money to secure the project, site control, good to go. They apply to the OPA
> [Ontario Power Authority] and claim to be a community power project,
> and the community says no. Meanwhile, you get the adder for this group
> claiming to be community. So we decided that it means at least 10 percent
> local equity, for at least a stake in the game. The other 41 percent to get the
> adder could come from a nonlocal community power. (Personal interview,
> July 24, 2009)

Sometimes the definition includes independent power producers and financial institutions (OSEA 2011), sometimes farmers (Government of Ontario 2009), and sometimes none of these (Nova Scotia Department of Energy 2010). In Nova Scotia, Community Economic Development Investment Funds are included as well. For example, according to the World Wind Energy Association Community Power Working Group:

A project can be defined as Community Power if at least two of the following three criteria are fulfilled:

Local stakeholders own the majority or all of a project:

A local individual or a group of local stakeholders, whether they are farmers, co-operatives, independent power producers, financial institutions, municipalities, schools, etc., own, immediately or eventually, the majority or all of a project.

Voting control rests with the community-based organization:

The community-based organization made up of local stakeholders has the majority of the voting rights concerning the decisions taken on the project.

The majority of social and economic benefits are distributed locally:

The major part or all of the social and economic benefits are returned to the local community. (OSEA 2011)

This inclusion here of financial institutions and independent power producers (IPPs) differs somewhat from the Ontario Power Authority's rules for awarding community feed-in tariff adders:

Community Investment Members means,

one or more individuals Resident in Ontario;
a Registered Charity with its head office in Ontario;
a Not-For-Profit Organization with its head office in Ontario; or
a "co-operative corporation," as defined in the Co-operative Corporations Act (Ontario), all of whose members are Resident in Ontario. (OPA 2013a)

Which actors are included in the definition of "community" *matters*. This is because these projects play a key role in legitimating renewable energy projects, helping overcome local opposition to development sites. This local involvement is particularly important in jurisdictions where there is a strong suspicion of the mechanisms through which renewable sources are being developed. This is, in fact, one of the reasons community power policies are heralded as an important policy mechanism for developing new renewables. Some actors argue that electricity co-operatives are central and necessary, whereas others argue that they are a tool and part of a mixed community power sector where (for-profit) private actors need to play a role, at least initially (personal interviews, July 20, July 24, and October 14, 2009).

NIMBYism and Development Opposition

Another political challenge arising for co-operatives in the community power sector is the relationship between community projects and local project opposition. Electricity generation projects have environmental consequences. They have resource impacts, livelihood impacts, and a whole host of implications for local communities. Community partnership structures are increasingly likely as large industrial wind farms account for a larger share of electricity generation in Canada. These local partnerships may help the projects move faster through project approval processes. But caution needs to be exercised here, lest the important institutional links between community participation and power be severed.

What is unclear in debates over new renewables development is whether and when the NIMBY pejorative is appropriate. Sometimes this pits co-op developers against other locals. The benefit of community power therefore turns into a shaming of local project opponents. The power seems to be with pro-project and pro-development forces, whereas those opposing are cast as obstructionist NIMBYs. Community power needs to include the local control to say no: there needs to be real consideration of whether the opposition is to the new source or to the lack of democratic siting decisions and ownership of renewables. Community power developers wishing to remain anonymous have and continue to be approached by larger power developers to come in and "talk to the community" in order to secure local confidence in a project, even though very little actual interest exists in transferring ownership or power to local groups.

What has emerged is a split between renewables enthusiasts, de-growth deep greens, and NIMBYs in the sense of landowners who are opposed to noise and the aesthetic impacts. As I have argued throughout this book, one of the key advantages of community investment in renewable projects cited in the literature is that a stake in the project – where those bearing the physical costs of the project benefit in jobs, money, and control – decreases project opposition. In addition to attempts by private developers to co-opt community power experts, other challenges for communities have emerged.

In Nova Scotia, tension between a CEDIF (community investment) wind developer, Scotian Windfields, and residents of Digby emerged when the company proposed to site a twenty-turbine wind farm in the coastal community. Some residents were supportive and interested in investing, but others were vocally opposed. According to Janice Ashworth of the Ottawa Renewable Energy Co-operative, "it was mostly opposition from one family ... they are back-to-the-land and happy with wind turbines for their own consumption or community, but they're against industrialization of anything. So, [it was] 'Put us and our community off the grid; we'd be fine with wind and solar, we just don't want to support pulp plants and industrialization'" (Ashworth personal interview, May 21, 2010). Although it is unrealistic to expect new developments to be unanimously supported, these issues do point to clear political tensions in decisions about who has control, who speaks for communities, and how community power is both understood and used going forward. It also raises questions about how limits and consumption levels are bracketed off from energy-sector reforms. The Digby project was purchased by Nova Scotia Power (Emera) in February 2010 and became operational in December 2010.

This discussion about the importance of overcoming NIMBYism and moving projects forward raises several issues. Barry, Ellis, and Robinson (2008) explored the role of community opposition, paying particular attention to the rhetorical constructions surrounding the term "NIMBYism," in the United Kingdom. They found that although an element of climate change denial and conventional NIMBYism exists in local opposition movements to wind farms, there is also a strong suspicion of the mechanisms through which renewable sources are being developed. One concern is that utility companies are making money at the community's (and public's) expense. They found that the real basis for skepticism over renewables was, in fact, a lack of trust in government, regulatory processes, and wind farm developers. For Barry, Ellis, and Robinson (2008, 82), "those presenting the anti-wind energy position are keen not to be regarded as motivated by self-interest, but are skeptical of 'non-local forces' (state and business) coming in and trying to pull the wool over their eyes with what they see as 'PR stunts' portrayed as consultations."

These arguments based on the UK context suggest that overcoming opposition to renewables development is not just a matter of more information for a misguided populace, though in some cases that may prove to be the case. The key in developing renewables democratically is in actually engaging local people in the development of and profits from projects – that is, real community power (MacArthur 2015). It requires a strengthened public-policy intervention to restrain the overwhelming dominance of other private actors in developing renewables for profit rather than community use. In Canada, then, overcoming community resistance to project development needs to be considered in light of these findings and the actual reasons for opposition examined. There is awareness within the community power and co-operative sector of the issues with development. Indeed, the recommendations for a round two of the Ontario FIT include clearer set-asides just for communities, and more supports to ensure that the partnership shares that communities get are maximized.

Summary

Networks built around and including electricity co-operatives have emerged with significant impacts. This is particularly so in Ontario's community power sector. The value of new co-operative developments thus goes beyond monetary gain to a transformative role for projects in shaping public opinions, experiences, and, through that, policy. The interactive role between the constituencies created by community groups and policy change is well documented in the Danish case, but it is also evident in Canada. For example, WindShare Co-operative, in Toronto, started a coalition and created momentum toward what is now the Green Energy Act. Alliances are forming – not always easily – between co-operatives, some private developers, municipalities, and First Nations to advocate for renewables policies that support local participation in generation projects. However, significant political and political economy challenges remain. The empirical reality remains that despite all of these networks, co-operatives represent a tiny share of the larger privatizing renewables sector. Clear and important tensions have emerged over the definition of "community power" and the use of community power as a legitimating tool for private-sector renewable development.

9

Empowering Electricity

It is an exciting time to be following electricity co-operative development in Canada. Nearly a decade of research has yielded more depth and diversity in these organizations than I could have imagined. In 2006, no Canadian provinces had policy supports targeted for electricity co-operative development. WindShare's one turbine was the only co-operative wind-generation project in the country, and it seemed as if the task of tracking and assessing its progress would be relatively simple. How times have changed. As of 2015, at least 20 co-operatives are generating power on more than 100 projects. There are more than 202 electricity co-operatives of all types operating across Canada. These co-operatives function across multiple areas: generation, distribution, retail, and education. The decisions being made to start them centre overwhelmingly on helping communities address coming challenges in the electricity sector: rising prices, concern over nonrenewable fuel sources, and lack of participatory control. Their focus is on re-embedding the power sector in local ownership and local networks, albeit with success currently limited to small pockets of the country and requiring significant public-policy support.

The development of these organizations is crucially important today given the scope and scale of the climate crisis. Intergovernmental Panel on Climate Change scientists have grown increasingly emphatic about the pressing nature of the climate challenge and its significant impacts on food security, economic growth, and species extinction (Evans 2014; IPCC 2007,

2014). It will become more difficult as time passes to stabilize GHG concentrations, so a policy emphasis focusing on adaptation rather than mitigation is clearly unwise. Some new developments are promising: from 2004 to 2012, global renewable power capacity for solar, wind, tidal, biomass, and hydro sources nearly doubled from 800 GW to 1,560 GW; it increased 8 percent from 2012 alone, and in 2013, more solar photovoltaics than wind capacity was added for the first time in history (REN21 2014, 15). However, renewable electricity's share of overall global generation is increasing slowly because overall demand is rising rapidly and new renewables are more likely to be variable rather than baseload power sources (IPCC 2011). It is clear that governance transformations are required to solve these problems; we need to change not only what we produce and consume but also how and why and where we do it.

I began this book with numerous questions about whether electricity co-operatives formed part of a Polanyian double movement against the "dangerous fiction" of free-market neoliberalism and constituted the basis of a democratized power sector. This chapter draws together core insights from the preceding pages and presents directions for future research on this topic. It links electricity-sector policies to the discussion of crises and the double movement in Chapters 1 and 2. I illustrate how and where the co-operatives outlined in this book fit within an empowered participatory governance model, and where substantial gaps and challenges remain. One of these gaps in understanding is the relationship between the significant – and unique among member countries of the Organisation for Economic Co-operation and Development – public electricity systems across Canada, electricity co-operative development, and environmental sustainability. This book has, in many ways, only scratched the surface of examining the contribution these organizations can and should make in Canadian electricity systems. The final section sketches four important lines of research going forward.

Power Policy Gaps, Crises, and the Double Movement

Electricity co-operatives are certainly a response to neoliberal developments: they are driven, in part, by the fact that Canadian energy policies on the whole have not been particularly bold or transformative. This manifests in the power sector because of a disjuncture between public opinion, needs, and public policies at the federal and provincial levels. At municipal and community levels, there is an impetus for collective action, particularly in provincial electricity regimes with significant private power ownership and heavy reliance on coal as a source of generation. Although electricity regimes are provincial, federal retreats both from environmental policies and public ownership impact communities in important ways (Bratt 2012; Gattinger and Hale 2010; Harrison 2012). The partial and uneven progress of power-sector restructuring across the country creates a different policy context and contribution of electricity co-operatives in each province.

The oil price spikes in both 2008 and 2010, as well as the contribution of fossil fuel–based electricity generation to global warming, have made reforms supporting new renewables generation more acceptable politically to the Canadian population. According to a recent Environics research poll, "over 70 percent of Canadians agreed that money spent on wars and the military would all be better spent on efforts that reduce greenhouse gas emissions and the impacts of climate change"; as well, "over 80 percent of Canadians believe the Canadian government should invest in 'green jobs' and transition programs for workers and communities negatively affected by a shift off of fossil fuels" (Penner 2011). This public support in principle for sustainability initiatives failed to translate into meaningful political action or public policy supporting deep sustainability under the Conservative government of Stephen Harper. For example, in December 2011, after years of inaction and support for the oil and gas industries in Canada, the Conservative government formally withdrew from the Kyoto Protocol (Curry and McCarthy 2011). For seven years running Canada received the Fossil of the Year award from international environmental groups (Climate Action Network 2011) and has grown increasingly isolated at the international level. In addition, the downsizing of environmental ministries (Campion-Smith 2011) and the silencing of federal environmental researchers in the late 2000s sent the signal that federal policy makers were unlikely to take significant action to curb greenhouse gas emissions. With the election of a new federal government in October 2015, there are hopes this policy orientation will change, but it is too early to tell exactly where and how new initiatives will contribute to reorienting Canadian infrastructure toward a post-carbon future.

These policy failures at the federal level have resulted in pressure on subnational levels to drive change, and on a search for new mechanisms of influence and innovation. Certainly in Ontario, the Green Energy and Green Economy Act, with its community adder, has gone furthest to support both community power and reduce fossil fuel power generation. This was enacted subsequent to private power restructuring in that province and the failure of earlier policies to generate new renewables developments (Chapter 5). The challenge is not that Canadians need to want greener power; it is that the neoliberal rhetoric surrounding free choice and positive-sum outcomes makes greening about increasing profitability rather than ecological necessity. According to Robinson (2007, 15), the drawback of the market-based environmental approach is that "it effectively rules out confronting the issues of how energy is produced, on the one hand, and consumed, on the other. It is easy to understand why we have witnessed two decades of climate change policies that were in effect predestined to fail." Neoliberal green policy is driving policies supporting green consumption and green business, and undermining the fundamental point of social ecologists: that deep sustainability requires systemic change and a shift away from ever-increasing material consumption and the prioritization of profit as a central feature of economic systems.

The specific policy mechanisms employed to green the power sector have long-term political economy impacts, which in turn will shape their effectiveness. Moves to address environmental degradation have become highly politicized, as they are intimately connected with the relative economic power of different industries and business versus citizens (Johnston, Gismondi, and Goodman 2006b). As prices for power rise – which indeed they will with the new investments required – private ownership of these projects shifts resources (power) and rents to private investors rather than to public agencies – a process of greenshifting. This has distributive consequences for the types and allocation of jobs, of who can – and does – pay specific actors various rates for usage. When private investment plays a central role in the process, public agencies cede the rents from projects, and thus the financial tools to facilitate deeper greening projects (such as demand management and incentives for efficiency building). For community actors, then, the Faustian choice was to leave renewables to the private sector or push for a share, however small, of that emergent market. When markets are created for private power development, co-operatives secure only a tiny portion (if any) of new contracts, without significant pressure from social movements.

Institutions that arise from within capitalist processes, such as co-operatives, may form part of a transformational response to the dislocations and contradictions within the current system. Without control of resources and methods and mechanisms of production more broadly, social and environmental externalities will continue to erode the security of citizens. The nature of the reforms taking place in the electricity sector in Canada is thus at odds with the holistic systemic overhaul required to deal with the triple crisis. Electricity co-operatives occupy a contradictory position in that they are both a product of a desire for deep systemic change and a product of policies that deepen the crisis.

New electricity co-operatives are a response to a lack of local project control in the power sector. Co-operatives emerging today – particularly nonprofit, solidarity, and worker ones – are part of the broader global justice move represented in alternative globalization movements (Cavanagh and Mander 2005; Neamtan 2002; Sklair 2002). For example, the move-your-money initiative to shift funds to credit unions and recent Occupy protests both featured co-operative and collective associations prominently as part of a range of existing alternatives. They reflect a desire for social and local control over human and natural resources. The interviewees for this book overwhelmingly argued that the motivation driving new co-operative formation was to retake and reform electricity systems, and to re-embed resource control locally after private actors have entered the sector. They are, often explicitly, mobilizing people – frequently with no experience or prior interest in co-operatives – to engage and change what they feel are problematic ownership and policy choices.

For co-operative members, the argument to retake, reclaim, and own their own generation, rather than cede opportunities to private developers, comes with an implicit critique of private power. Also embodied in this narrative is a critique of a particular model of public power, one in which elite technocrats in urban centres are wedded to technologies and systems that impose disproportionate costs on (often rural) communities. The power of incumbent industry actors can, and has, led to "iron triangles" in policy making where fossil fuel and nuclear industries enjoy disproportionate power and influence (Bratt 2006; Durant 2009). This leads to charges that policy is hierarchical and technocentric, and lacks effective democratic processes (Devine-Wright 2011; Etcheverry 2013; Harden-Donahue and Peart 2009).

There are logical reasons to desire a shift toward co-operative and community power that have little to do with neoliberal orthodoxy or continental aspirations. As illustrated in Chapters 1 and 2, close institutional connections between owners, users, and producers are increasingly important across many sectors (Ostrom 1990; Pateman 1988; Wright 2010). Disembedded private actors have no internal incentive to conserve or to reduce demand. Indeed, they face the opposite pressure, because once a resource is exhausted, they can relocate to another site. When private and nonlocal actors are introduced as resource managers and are regulated by public entities ideologically committed to market-based and industry-led regulation, the worst of both worlds for environmentally sustainable governance results. There are benefits to having local people trained, to having siting directly decided by people who live in and use the community, but persistent issues of equity and who gets to be in the community are crucial (Musall and Kuik 2011).

Local participation in resource development and management can facilitate economic development and capacity building, and deepen important knowledge bases. State agencies and firms may have access to resource data, but the public is often in the dark about the real value of these resources. This basic ignorance is problematic, since communities are often "asleep" when valuable land is leased, and resources are sold off or exploited (Walker 2008). Local share in profits is thus left to securing a job at the mine or plant, or via indirect benefits from property and municipal royalties and taxes. This is increasingly an issue for new renewables such as wind, just as it has always been with fossil fuel resources, and is directly affecting communities in Canada that are trying to start electricity co-operatives (phone interview, April 13, 2010; Ferrari personal interview, July 23, 2009; Gagnon personal interview, May 16, 2010). These social and political failures, when set in a global context of austerity and economic contraction and a continental context of tightening energy co-operation, mean that revolutionary change, rather than piecemeal and partial sectoral reform, is essential. For that to take place, we need to take seriously the many theorists urging multiscalar collective action and those cautioning against a fetishization of the local (Albo 2006; Catney, MacGregor, et al. 2013; De Young and Princen 2012; Johnston, Gismondi, and Goodman 2006a).

The "Electricity Co-operative Difference" and Empowered Participatory Governance

The electricity sector is clearly just one in a wide range of places where a shift in governance from concentrated and hierarchical to more diffuse, democratic, and empowered needs to take place. In the literature on community renewables, much is made of the improved distribution of costs and benefits in a community-based electricity system: it's more egalitarian, more democratic, more effective at reducing local opposition. The degree to which these advantages materialize in practice is influenced to a large degree by how open the project-planning processes are, as well as by how collective and locally based the project is. There is, however, no easy link between one type of organizational form (co-operative, nonprofit society, and so on) and the degree to which a particular project meets strong conditions for community benefit and participation. Much depends on the opportunities provided by the regulations and policy supports, political cultures, and the ethos of the mobilizing group and their commitment to either private economic development or broader public benefit.

Co-operative electricity initiatives can indeed develop a capacity for provoking and deepening sustainable transitions (see Chapters 7 and 8), particularly if they are scaled up to become a central rather than a marginal part of otherwise private power sectors. Significant scale-up may have begun in Ontario, but steep challenges remain there. Hence, in 2016, there is some potential for, but little practice of, empowered participatory governance (EPG) in Canada. Electricity co-operatives are, however, (1) impacting policy, (2) scaling up to larger networks and systems, (3) extending into highly competitive and technical sectors, and (4) raising significant capital. Deepening the co-operative role in EPG rests on co-op support for public electricity regimes *and* broader social movements across the country.

Earlier in this book, Table 2.2 set out the evaluative criteria for the strength of co-operative electricity's contribution to EPG. Below I outline the results of how Canadian electricity co-operative developments, in general, fit within this structure. The five elements of empowered participatory governance are policy impact, education, asset ownership and control, networks, and finally, participatory and anticapitalist norms.

The Toronto Renewable Energy Co-operative and the WindShare Co-operative, for example, embodied several of these various contributions to EPG. The first comes from securing local control of resources so that their development is managed in sustainable ways and can be used for local economic development. The second is the role social economy actors play in modelling the possible and in pushing new technologies, management methods, or institutional forms. Third, social economy resource initiatives play a key role in combatting NIMBYism by engaging community members and giving them a stake in resource projects. Finally, these initiatives contribute, whether successful in operationalizing projects or not, to developing more informed, aware, and mobilized constituencies. These constituencies potentially mobilize and agitate for broader transformative policy and political change. Of course, the difference between the practice of one particular project and the potential of the broader organizational form and sector is dependent in a very significant way on the political economic context both across sectors and internationally.

Policy Impacts

As with co-operative rural electrification in Alberta and Quebec, specific policy drivers are playing an important role in both facilitating and constraining current co-operative development. In Alberta, rural electrification associations (REAs) were initially limited to serving only farms, were not granted franchise areas – as REAs in the United States have, or gas co-ops in Alberta have – and are prohibited from selling the REA to other co-ops. In Alberta and Quebec, governments keen to avoid the public option of electricity development taking place in Ontario facilitated co-operatives. This was despite rural areas being underserved. At present, new distribution co-operatives are not developing but generation co-operatives are. These new co-ops are facilitated by provinces looking outside the public sector (where there is a public utility) for new renewables development. Co-operative and community actors, however, are constrained by regulations to particular sizes and grid connections (FCPC 2013, 2014).

Electricity co-operatives are both driven by and driving policy. In Chapter 8, I illustrated the policy networks that have arisen out of electricity co-operative development and outlined where, when, and how this has spurred specific policy initiatives. In Nova Scotia (2010), New Brunswick (2010), Ontario (2009), and Quebec (2010), a range of policy supports for co-operative and community power generation were implemented. These arose out of pressure from community groups after private generation and contracting for new renewables development was already underway, and when they began to confront barriers from a tilted playing field. In all provinces, the general understanding gained from policy and community interviewees is that the co-operative and community contribution to power generation will be small in practice, despite some actors in the community power sectors advocating for 100 percent community ownership (e.g., OSEA). In some cases, these expectations of minimal development have arisen out of the very real constraints communities face when developing expensive projects in competition with other private developers: financing, grid access, site access, unequal lobbying power, and legal recognition.

Despite the challenges with particular policies, the fact remains that co-operatives have been somewhat successful, via their political mobilization and lobbying, in advancing their interests. In Alberta, REAs were able to modify legislation in order to serve nonfarm customers, and to self-operate retail and distribution without incorporating each business area separately (as other private actors have to). In Ontario, co-operatives, as part of a community power network, were successful in pushing changes to the co-operatives act that allowed for the incorporation of renewable energy co-operatives that sell electricity to the grid rather than to their members. These and other examples given in Chapters 6 to 8 illustrate that although co-operatives are by no means in a position to dictate terms to policy makers, they have been successful in particular instances.

Education and Public Engagement

A second aspect of the contribution co-operatives make to EPG is in increasing both public awareness about renewables and participation in resource ownership and management. Some electricity co-operatives (see Chapters 6 and 7) incorporated solely for the purpose of supporting and facilitating public education and policy change away from fossil fuel–based power and toward conservation and local accountability. Other co-operatives – the Toronto Renewable Energy Co-operative and Peace Energy Co-operative, for example – work on developing then spinning off generation co-operatives, as well as on public education, lobbying, and networking. In order to build EPG, broad engagement by the public with these in-practice alternatives is crucial. Participation and education varies between types of electricity co-operatives, but all co-operatives studied in this research evidenced at least a minimal commitment to the co-operative difference, community education, and local development.

Participation in the development of new renewables is vital for project success and local development. Effective public engagement – not to be confused with consultation – is a central challenge for new initiatives in the energy sector. The displacement effects of dams and power plants can be mitigated, in part, by allowing affected publics to play a greater part in the design, scale, and siting of projects. In Copenhagen, the Middelgrunden wind farm was, through extensive public participation, modified to fit with the natural curve of the bay (Pahl 2007). This move made the project not just more aesthetically appealing and a tourist attraction but locally supported rather than opposed. Since all viable power options are accompanied by physical and financial impacts, multilevel governance mechanisms that go well beyond thin consultation are essential to environmental sustainability (Andersson and Ostrom 2008; Turnbull 2007).

Ownership and Control

As with the co-operative sector more broadly, there is a significant variation in co-operative structures between specific projects and communities (see Chapter 7), and between consumer retail, generation, distribution, and educational forms. Some electricity co-operatives are participatory and movement based, whereas others are used as community investment vehicles rather than for actual project employment or management. Some own a more significant percentage of the project than others (SolarShare compared with PEC's Bear Mountain, for example). The diversity in projects is a function of the range of challenges facing communities interested in co-operative development. These challenges include difficulty raising the significant funds required for generation projects, grid connection lines tied up by other private developers, challenges with legal incorporation, and pressure to sell distribution co-operative assets to private utilities.

The challenges for co-ops arise from lack of familiarity and support for community-owned and co-operative models coupled with significant asymmetries in power, resources, and information between community developers and other private companies. Sectoral competition pushes co-operatives to form partnerships with private developers. Where community power development becomes little more than consultation and communitywash, a problem emerges. It waters down the co-operative difference, the ultimate economic benefit, and local control, while legitimating the new private development because there is a portion of community buy-in. The process of communitywashing is complex, but it does not lead to the development of empowered participatory governance in the electricity sector. For that, community and public ownership need to become much more than a small segment of a broader privatizing power market.

If communities are going to be able to scale up in the power sector, they need to have first access to local development sites. Co-operatives often have a longer learning curve and, being volunteer based, are less able to dedicate the time and resources that private power companies can to sites. In addition, many of the financial challenges could be overcome with policy and program supports. Some of those suggested include government loan guarantees, RRSP eligibility for community power, financing, and infrastructure investments. Financial incentives are not enough, however, to ensure that co-operative electricity projects succeed in deepening EPG, as long as the broader power sector is oriented toward increased generation for export and profit.

The Power of Networks

Co-operative networks are a strength of the movement, both in Canada and internationally (see Chapter 3). In the electricity sector, co-operatives are also deeply networked not only within provincial co-operative associations but also with renewable energy associations, within the social economy, and with international renewables and community power movements. This has implications for EPG insofar as it signifies great potential to leverage the positive experiences of different actors and jurisdictions, and to learn from their challenges and failures. For example, research from the United Kingdom suggests that community electricity can be, and in fact is being, hollowed out of its promise by the larger progress toward partnerships and power for profit. On the other hand, drawing from the experience within Nordfriesland, in Germany, one model that could serve to deepen the prospects for EPG project development is tiered power calls wherein the first access for sites go to those in a local community, then to the larger region, province, and finally beyond. This means that those more directly affected by the development have a chance to participate through investment and help guide decisions about the project through, for instance, co-operative member meetings.

In the community power networks developing in Canada, co-operatives are but one of a number of municipal, First Nations, nonprofit, and private actors, with a range of ideas about what the movement's ultimate goals are. These identities are still being negotiated in different provincial jurisdictions, and many actors within the co-operative movement are aware of the complexities and challenges of co-optation, which may bode well for future developments. That said, the very fact that multiple networks across multiple boundaries are developing and strengthening suggests that a long-term movement with scale-up potential is being built. This networking activity bolsters claims that electricity co-operatives are part of a broader movement with some strength, rather than simply a small niche with little power.

Anticapitalism and Participatory Democracy

Finally, the norms animating emergent co-operative and community power movements matter. They shape the direction of ultimate goals as well as set the boundaries for the kind of compromises and coalitions that actors are willing to engage in (Golob, Podnar, and Lah 2009). Anticapitalist and participatory democratic norms were identified as representing a significant divergence from those underpinning mainstream neoliberal governance. Although nearly all co-operative members interviewed for this research outlined the co-operative difference by referring to participatory democracy; one member, one vote principles; and the need for economic democracy, explicitly anticapitalist sentiments were rare. This is, perhaps, unsurprising given the history of the co-operative sector as a "third way," and social enterprises as capitalism with a human face.

In Chapter 2, I argued that newer co-operatives may – given recent historical developments toward solidarity economies and antiglobalization movements – have evidenced more explicitly political orientations. This I did not find. What emerged from the research, however, was a frustration with a mainstream political economy in which newly created markets were far from free, large companies dominate the power sector, and both rural economies and urban centres were heading in unsustainable directions. In fact, within new electricity co-operatives, one of the major drivers for new members was an environmental shift toward the development of new renewables as well as energy efficiency. This has interesting political implications: it made the members interviewed for this book far less likely to support Conservative policies, compared with co-operative members in other areas of the energy sector (gas, petroleum refining, and retail).

Ultimately, although the normative commitments within electricity co-operatives emphasized participation and democratic decision making, the practice did not necessarily follow. Members across the range of electricity co-operatives reported initially high turnouts for general meetings, but as projects progressed, turnout diminished. This is unsurprising given the time constraints facing Canadians. Many co-operatives worked to produce newsletters (sent by mail or available online) in addition to holding formal meetings. They did this to encourage active participation and meeting attendance, and, at a minimum, to keep members informed. It is thus important to recognize that although the co-operative form provides the space and the promise of project participation, it does not always translate into significant participatory democratic decision making.

Electricity Co-operatives, Sustainability, and Public Power

The intersections between the development of new co-operatives and the role they play in legitimating private power in provincial electricity regimes are problematic. Empowered participatory governance requires local control, democratic decision-making structures, and networked countervailing power. Environmental sustainability in the power sector requires a strong state capable of hard emissions caps, redistribution of costs, and intervention in powerful polluting industries. Taken together, this means that sustainable

EPG requires a strong public sector *and* significant local control: public and participatory green power. Unfortunately, these political economy arrangements are radically different from contemporary neoliberal shifts taking place in Canada (see Chapters 5 and 7), wherein the power and prevalence of private renewables development drives Canadian electricity sectors in a very different direction.

Bold renewables policies focused on public engagement, transparency, and conservation are necessary for the creation of a sustainable electricity sector. This is vitally important because public backlash against environmental reforms is one consequence of divorcing environmental policies from the economic realities of a given population. In Quebec, many communities are firmly opposed to wind-power developments (Gagnon personal interviews, May 5, 2010; May 16, 2010). In British Columbia, a carbon tax – seen as a key element in greening the power sector by most environmentalists – was hugely unpopular, as were some aspects of the Green Energy Act in Ontario. Conceptualizing climate change and environmental degradation as a fundamentally human problem of distribution and justice requires an active and interventionist state. Current incentives and targets are ineffective in most provinces (Hoberg and Rowlands 2012; Murphy and Murphy 2012); in Ontario, which has taken the boldest steps of all the provinces to support community actors, far more work is still needed (FCPC 2013; Gipe 2013; Stokes 2013b).

A committed change starts with modifying prices so that they take into account externalities (Perkins 1998; Princen 2001) and requires significant coordination across provinces, states, and issue areas. These environmental challenges, particularly those around scale and governance, make co-operatives and community power look like an important and potentially powerful direction, or at the very least one that needs more attention. This is especially the case when one considers that these organizations are playing pioneering and innovative roles with newer technologies; they are leading the way in solar, tidal district heating, and other developments because of their lower profit requirements (Bourne personal interview, December 1, 2009; Cumbers 2013; ILO 2013).

Public power, however, is the real elephant in the room in the community power sector. In Canada, unlike in the United States, the United Kingdom, or Australia, public ownership still characterizes the generation of electricity in most jurisdictions (Alberta, Ontario, and Nova Scotia excepted). Public-sector governance of electricity structured around public utilities has been criticized for problematic relationships with large industry and centralization of power, as well as its lack of local involvement in siting and approvals decisions. These challenges do not disappear when markets take over (see Chapter 4). I have no illusions about the environmental consequences of large dam developments, but it is crucial to point out at this turning point in Canadian power-sector history that the public option has, in many provinces, been relatively green. But it has not been particularly participatory or conducive to the kinds of restructuring required in the future, so transformation is clearly necessary. Put another way, ownership matters, but so too do the mechanisms through which policy decisions are made (Fung and Wright 2003; Keevers, Skykes, and Treleaven 2008; Newig and Fritsch 2009).

If we accept, as many seem to today, that green power means private power, then the choice to increase the community and local content seems simple. However, the definitional fiat equating private power with innovation and renewables is, in Canada, ahistorical, inaccurate, and ultimately dangerous for eco-social sustainability. A significant challenge is thus raised by new co-operative generation projects over the ultimate impact of provincial electricity regimes. Not all provinces' electricity systems are coal based, nor are private actors the incumbent generator; most are public and hydro based. In Manitoba, Quebec, and British Columbia, co-operatives become part of a rollback of public ownership for new renewables. That means a shift to power generation for shareholder profit, and in some provinces for export, rather than for domestic and local needs. Canadian provinces are at a crucial juncture wherein the continental and international pressures to reduce the debt loads of public agencies place significant pressure on public ownership (see Chapter 5). In times of global and national austerity, power assets are often among the first on the auction block (for example, Ireland and Greece in 2011). The public-private power debate matters in a more significant way in Canada than in countries with already restructured power systems. Co-operative actors here need to be clear about the kind of new alternative they can form, given that in restructured power markets without significant public regulatory and financial support, they are outbid and outmanoeuvred by other private actors.

These broader issues of political economy certainly affect the ultimate aggregate contribution co-ops can make toward sustainability in the electricity sector. Electricity co-operatives are not driving the breakdown of public power. They are, however, putting a human face on the private power sector by redirecting some of the profits back into places that would have far less investment or control without them. People formed co-operatives after public policies opened space for private power and scaled back or abandoned public utility development, first in distribution and now in generation. If the goal is democratization and greening of the power sector, recognition of existing provincial systems is vital, and these differ across the country. We need to be cognizant of whether new generation projects are actually needed. Environmental sustainability requires minimizing throughput of materials, minimizing the power we collectively use and generate.

Thus, prospects for a net positive electricity-co-operative contribution to sustainable empowered participatory governance are maximized in provinces that are coal reliant and already have a majority of private-sector ownership of power generation, like Alberta and Nova Scotia. In public fossil fuel–based provinces like Saskatchewan and New Brunswick, new co-operatives could best play a role in partnerships with public agencies in order to maximize the local benefits of co-operativism *together* with the fundraising supports and community connections of the co-operative. This is not to suggest that such a partnership will be either natural or easy: bureaucracies are slow to change. However, if the goal is transforming energy systems in a significant way, state-level transformations are required as well as individual-level ones (Barry 2012; Eckersley 2004). In Quebec, Manitoba, and British Columbia, where existing public large-scale hydropower is the primary generation source, contradictions arise with co-operatives as independent power producers opening the door to profit-based power development.

Without broader intervention and ownership by the public sector, there is a danger to sustainable community development. The libertarian and strong communitarian instincts that encourage "self-help" result in some communities (or members within particular communities) being left behind. Communities and citizens that have the money to invest in electricity generation benefit from lower rates and, in some provinces, greener power. For the rest of the people in the province, electricity rates are climbing, and social transfers are not offsetting these costs. These increased costs are certainly not solely attributable to new renewable generation – costs of nuclear refurbishment and subsidies play a key role in Ontario – but they do significantly affect the broader population. For those who don't have $1,000 for the solar investment, or money for a share purchase in the turbine, their role in private power markets is reduced to that of a disempowered consumer. Private generation is not the only alternative to coal-fired or nuclear power. On the one hand, the more co-operative organizations are developed and running, the more resilience communities will have to keep themselves going when crises hit. On the other hand, a problem arises if co-operative actors are participating (actively or tacitly) in the breakdown of public ownership. Co-operatives differ in important ways from other private actors, but they are not public. Members, rather than the whole community, own resources in these new projects, and redistributive links between a wide set of stakeholders are not always developed.

Fuel poverty is an increasing reality for poorer Canadians as prices rise, forcing them to choose between heating and food. If innovations bringing decisions closer to users need to take place (as I would argue they do), reforms that devolve power and responsibility for distributed generation closer to municipal levels, perhaps in co-operation with co-operative community investment projects, would be useful. A model we should examine more closely is one that melds public and community, as in Denmark, to draw on the benefits of both through co-operative-municipal partnerships. In that way, the participatory governance benefits are harnessed, as are the institutional supports and redistributive benefits of comprehensive (rather than solely member-based) public control. This has the advantage of retaining the public institutional experience and financing leverage, together with the participatory movement-based mobilization of the co-operative.

Directions for Future Research

The conclusions drawn from my research suggest several fruitful avenues for future study of the role of electricity co-operatives in Canada. As these organizations continue to develop, as new electricity policies are implemented, and as the processes of environmental degradation and climate change continue, the importance of this understanding grows. The research presented in this book was constrained by my choice of such a broad – both geographically and conceptually – and understudied area. As a result, a range of valuable projects have emerged that break down along four main lines: further international comparisons, co-operative supporting public-policy prospects within the context of international trade rules and regimes (NAFTA, WTO), case-based generation comparisons with within-in-province private power generators, and finally, case-based studies of public municipal- or provincial-co-operative electricity partnerships.

Comparative studies drawing from the experiences and policy contexts that supported American, Danish, and German electricity co-operatives would add a great deal to Canadian policy development in this area. Comparisons between electricity generation regimes in Denmark or Germany and one or two Canadian provinces would result in important data on how replicable Danish and German models are in Canada. Some work has begun down this line already, by actors in the community power sector and by academics working in England and the United States (e.g., Christianson 2011; B.K. Sovacool 2008; Warren and McFadyen 2010). A deeper focus on Canada, and particularly the political economy issues in specific provincial power sectors, is necessary, however, in order to use the German and Danish examples in a way that is actually meaningful in a Canadian context. This would include analysis of specific policies, along with the political coalitions required to see them to fruition, and where and how opposition was overcome (if it was). For example, Denmark, which generates 30 percent of its electricity from wind power, is connected to firm hydro resources in Norway and Sweden (NEB 2010a, 35). It also has the largest development of combined heat and power facilities in the world, which contribute to greater system stability and the ability to support increasing levels of variable power sources (Bouffard and Kirschen 2008; Hand et al. 2012). Understanding and acknowledging the enabling conditions beyond the new renewable and co-operative development is important.

The second area of extended research would probe more deeply into the ramifications of co-operative supportive policies raised throughout this book given the international trade regimes Canada is a part of (WTO, NAFTA). The trade rulings against Canada discussed in Chapter 5, together with a federal government committed to deepening trade liberalization, may forestall the kinds of policies required to promote co-operative success. Unlike in Germany and Denmark, where co-operative electricity generation has developed with some strength, the trade rules governing federal and provincial government procurement are subject to NAFTA. Since co-operatives are private-sector actors, incentives such as loan guarantees and privileged siting could conceivably be construed as a form of subsidy to Canadian corporations. Although provincial procurement favouring Canadian suppliers is protected under NAFTA Chapter 10 and the GATT at this point, these protections are being rolled back. The February 2010 Agreement on Government Procurement (AGP) between Canada and the United States extended national treatment to the provincial level. In December 2011, Canada signed the World Trade Organization's AGP, extending national treatment for provincial and territorial procurement (Annex 2) to other countries, with some important exemptions (WTO 2011).[1] What is important here is that the window of exempted goods, services, and actors under liberalized trade laws seems to be narrowing, particularly with the continuing negotiations over trade in services agreements which include power services (Lord 2011; Wilke 2011). More work is needed in order to understand how feasible co-operative supportive policy is as trade rules change, particularly with reference to electricity investments and provision of electric power. It is likely that as we move forward, policies privileging local co-operative development, should they result in any great success, would be subject to challenge.

A third avenue of future research could involve case-based comparisons of electricity generation co-operatives with similar (in source and provincial electricity regime) private power developers. For example, one study widely cited in the community power literature compared the local economic multiplier between local, in-state, and out-of-state ownership of wind-generation projects (Galuzzo 2005). Similar empirical work in different Canadian provinces comparing local ownership and outside ownership would make a useful supplement to my research presented here. A comparison between a range of ownership types – private nonlocal, municipal, and co-operative – would be similarly useful. This could, for instance, compare the local economic benefits of private run-of-river, wind, or solar developments with similar co-operative projects. As more projects develop across the country, this comparative research becomes both necessary and feasible.

A fourth research area would investigate models, constraints, and prospects for co-operative–public-sector partnerships, as well as for co-operative First Nations partnerships. As this research progressed, the co-op–municipal partnership model emerged as one possible way to draw on the institutional strengths of both actors. The TREC-WindShare model is but one example of this, and it is important to understand whether these types of partnerships are feasible across a number of jurisdictions and what their limitations might be. Identifying the limitations, as well as the specific results of partnerships that emerged from projects, would contribute much to our understanding of the possibility for scale-up and co-operative–public-sector co-operation.

Summary

The organizational and policy experiments outlined in this book illustrate that political support exists in some areas for prioritization of community power. They also illustrate that there are a range of Canadians seeking more democratic alternative structures in the economy, even to the point where they are willing to forgo economic returns on their investments in order to prove that projects can be built and help educate their neighbours and their children. I have great respect for the practitioners on the ground, in communities across the country, frustrated with what they consider a fundamental erosion of control over the most important resources and levers of power – in both senses – in Canadian society. As a result, I have attempted to analyze the development of electricity co-operatives in a way that is sensitive to the motivations, the volunteer efforts, and the long and often exciting history of co-operativism.

The rising profile of co-operatives and the social economy in many countries around the world suggests that we may be seeing many more of these organizations in the future. Co-operative ownership structures, even in technologically complex and financially challenging sectors like electricity, are both possible and desirable, despite their challenges. When contradictions arise between the direction of public policy and the values and opinions of members of the public, the co-operative model is one mechanism through which collective action and mobilization can be pursued. However, significant economic and political challenges confront the community power sector. Scaling up into an empowering and sustainable Canadian power alternative is unlikely without a strong public power movement that includes provincial, municipal, First Nations, and co-operative ownership.

Contemporary forms of greenwashing that generate temporary affluence as part of a new business opportunity – and do not address root causes of instability, environmental degradation, and exploitation – are problematic. Ultimately, even if some communities carve out a niche in this broader system, emissions will not fall far enough: the pipelines will still be built, the tar sands expanded. Small pockets of resilience will likely survive, but the broader battle will be lost. Co-operative actors need to pursue their projects with an understanding of whether they are legitimating a broader project of state restructuring, one that ultimately undercuts the very public agencies we will need going forward. This deep movement toward empowered participatory governance requires new, multilevel, and varied models: polycultures of dissent, as Shiva (1993) refers to them. These cannot be based on profit, ever-expanding growth, and business as usual. These transitions will take time and ultimately may not be successful given the array of challenges they face. But if they are, electricity co-operatives are well placed to play a role in this transition. For this to happen, it is crucial to strategically link these local movements to projects for systemic change – toward empowering power.

APPENDIX 1

List of Personal Communications

Interviewee	Affiliation[1]	Date	Type
Anonymous	Nova Scotia Department of Energy	May 19, 2010	In person
Anonymous	Windfarm developer, Nova Scotia	May 20, 2010	In person
Anonymous	Quebec co-operative developer	May 5, 2010	E-mail
Anonymous	Quebec co-operative researcher	May 13, 2010	In person
Anonymous	Nova Scotia Co-operative Council developer	May 18, 2010	In person
Anonymous	Ontario energy co-operative developer	April 1, 2010	Telephone
Anonymous	Former co-operative developer, provincial association	July 20, 2009	In person
Anonymous	Ontario wind co-operative board member	July 27, 2009	In person
Anonymous	Ontario energy co-operative developer	July 21, 2009	Telephone
Anonymous	Ontario renewable energy co-operative board member	July 20, 2009	In person
Anonymous	Ontario renewable energy co-operative member	July 23, 2009	In person
Anonymous	Alberta renewable energy co-operative developer	December 2, 2010	In person
Anonymous	Ontario energy association director	July 24, 2009	In person
Anonymous	Ontario co-operative developer	July 23, 2009	In person

Interviewee	Affiliation[1]	Date	Type
Anonymous	Ontario renewable energy co-operative director	July 23, 2009	In person
Anonymous	Ontario renewable energy co-operative developer	August 9, 2013	Telephone
Anonymous	National Energy Board	October 16, 2009	In person
Anonymous	Alberta Agriculture and Rural Development employee	November 27, 2009	In person
Anonymous	Alberta natural gas co-operative manager	December 1, 2010	In person
Anonymous	Rural Utilities Division, Alberta Department of Agriculture and Rural Development	November 27, 2009	In person
Anonymous	Director, Alberta natural gas co-operative; board member of REA	December 1, 2009	In person
Anonymous	First Nations renewable energy developer	April 13, 2009	Telephone
Anonymous	Pembina Institute researcher	November 27, 2009	In person
Alkalay, George	Owner, Northfield Ventures; co-operative consultant/developer	April 15, 2010	In person
Anderson, John	Director, Government Affairs and Public Policy, Canadian Co-operative Association	July 15, 2009	In person

▶

◄ **APPENDIX 1**

Interviewee	Affiliation[1]	Date	Type
Ashworth, Janice	Ecology Action Centre	May 21, 2010	In person
Atkinson, Pat	Saskatchewan MLA	July 9, 2009	In person
Barry, John	President, Seaforth Energy	May 19, 2010	In person
Bourne, Pat	General manager, CAREA	December 1, 2009	In person
Dahlstrom, Bud	Former CEO, Canadian Co-operative Refinery	July 13, 2009	In person
Dierker, Joseph Q.C.	McDougall Gauley	July 8, 2009	In person
Elkousi, Nada	Developer, Montreal-Laval Regional Development Co-operative (RDC)	May 14, 2010	In person
Empey, Harold	Federated Co-operatives	July 8, 2009	In person
Estabrooks, Bill	Nova Scotia energy minister	May 6, 2010	E-mail
Ferrari, Evan	Former president, WindShare Co-operative	July 23, 2009	In person
Flemming, David	Board member, Renew Co-op	May 21, 2010	Telephone
Gagnon, Martin	General director, Coopérative de développement régional, Bas-Saint-Laurent/Côte-Nord	May 16, 2010	In person
Gipe, Paul	Founder, Wind-Works; former acting director, OSEA	April 7, 2010	Telephone
Guénette-Lamontagne, Dominique	Conseil canadien de la coopération et de la mutualité	July 15, 2009	In person

▶

Interviewee	Affiliation[1]	Date	Type
Hebert, Yuill	Sustainability Solutions Co-operative	May 19, 2010	In person
Layton, Mike	Former deputy outreach director, Environmental Defence	July 27, 2009	In person
McLean, Joyce	Director, Strategic Issues, Toronto Hydro; former TREC board member	July 23, 2009	In person
Mole, Jeff	Bala Energy Co-operative	July 20, 2010	Telephone
Mowat, Jay	Everpure Biodiesel Co-op	July 20, 2009	In person
Nagel, Al	President, Alberta Federation of Rural Electrification Co-operatives	November 27, 2009	In person
Postle, Art	Former president, Federated Co-operatives	July 8, 2009	In person
Putt, Rhiannen	Formerly with Co-operatives Secretariat	July 16, 2009	In person
Regan, Don	COO, Municipal Electric Utilities of Nova Scotia Co-operative	May 21, 2010	Telephone
Rison, Bob	Former president, Peace Energy Co-operative	October 14, 2009	In person
Vincent, Bob	Former board member, Co-op Refinery Complex (formerly CCRL)	July 6, 2009	In person
Wyant, Peter	Crown Corporations Secretariat, Saskatchewan	July 13, 2009	In person

▶

◄ **APPENDIX 1**

Interviewee	Affiliation[1]	Date	Type
Zwicker, Barry	CEO, Scotian Windfields	May 20, 2010	In person

1 The affiliations listed here are at the date of the interviews. Please note that interviewees may have subsequently changed positions.

APPENDIX 2

Co-operative Policies and Programs across Canada

	Co-op–specific agencies/programs	Key legislation	Organizations[1]
Federal	Responsibility transferred from Agriculture and Agrifood to Industry Canada in 2013 as a Strategic Policy Sector Cooperative Development Initiative (2009–13)	Canadian Co-operatives Act 1999	Co-operatives and Mutuals Canada Canadian Co-operative Association Conseil canadien de la coopération et de la mutualité Canadian Worker Coop Federation Co-operative Housing Federation of Canada Credit Union Central of Canada
BC	No specific co-op agency	BC Co-operative Associations Act 1999 Amended in 2007 to allow for nonprofit and community service co-ops	BC Co-operative Association
AB	Rural Utilities Division, Department of Agriculture and Rural Development[2]	Alberta Co-operatives Act 2001 Includes specific provisions for new generation, multistakeholder, employment, and housing co-ops	Alberta Community and Co-operative Association
SK	No specific co-op agency[3]	Saskatchewan Co-operatives Act 1996 New Generation Co-operatives Act 1999	Saskatchewan Co-operative Association

▶

◄ **APPENDIX 2**   •

	Co-op–specific agencies/programs	Key legislation	Organizations[1]
MB	Housing and Community Development; Cooperative Development Services	Manitoba Co-operatives Act 1998 (updated June 17, 2010)	Manitoba Cooperative Association Conseil de développement économique des municipalités bilingues du Manitoba Conseil de la coopération du Manitoba
ON	No specific co-op agency	Ontario Co-operative Corporations Act 1990 Amended in 2010 to allow for renewable energy co-operatives to sell to nonmembers (the grid)	Ontario Co-operative Association Conseil de la coopération de l'Ontario (CCO) Ontario Worker Co-operative Federation
QC	Direction des coopératives Investissement Québec (administers the Co-operative Development Program, including the RRSP-eligible fund) (Régime d'investissement coopératif)	Quebec Cooperatives Act 1982 Nine updates, most recent in June 2011	Conseil québécois de la coopération et de la mutualité Fédération des coopératives de développement régional du Québec (with twelve regional CDRs) Co-operative federations by area: (forestry, worker, solidarity, etc.)
NS	Co-operatives Branch, Access Nova Scotia	Nova Scotia Co-operative Associations Act 1998 (amended 2001)	Nova Scotia Co-operative Council Conseil coopératif acadien de la Nouvelle-Écosse

▶

	Co-op–specific agencies/programs	Key legislation	Organizations[1]
NB	No specific co-op agency	New Brunswick Co-operative Associations Act 1978	Co-operative Enterprise Council of New Brunswick (CECNB)
NL	Newfoundland and Labrador Registry of Co-operatives Department of Innovation, Trade and Rural Development – Co-op Zone project and regional co-op developers network	Newfoundland Co-operatives Act 1998 (amended 2001)	Newfoundland-Labrador Federation of Co-ops (partner in Co-op Zone)
PEI	Registry of Co-operatives	PEI Co-operative Associations Act 1988 (updated 2009)	PEI Co-operative Council
NWT	No specific co-op agency	NWT Co-operative Associations Act 1988 (last amended 2006)	Arctic Co-operatives Limited
YT	No specific co-op agency	Yukon Cooperative Associations Act 2002	None
NU	No specific co-op agency	Nunavut Co-operative Associations Act (consolidation from NT, updated 2005)	Arctic Co-operatives Limited

1   There are also a number of sectorally specific organizations in most provinces – for example, the Federation of Community Power Co-operatives and the Alberta Federation of Rural Electrification Associations. Here I focus just on top-level organizations open to all co-operatives.
2   The Rural Utilities Division of the Department of Agriculture and Rural Development regulates the electricity and gas co-operatives in the province and has specific co-op expertise and links.
3   Prior to 2008, when the government formed Enterprise Saskatchewan, there was a Department of Regional Economic and Co-operative Development.
Sources: CCA 2011b; federal and provincial co-operative association websites in 2011.

APPENDIX 3

Top Ten Nonfinancial Co-ops, 2010

Co-op	Total revenues (million $)	Members	Employees (full/ part-time)	Activities
Federated Co-operatives (SK)	7,124	254	3,098	Wholesale, petroleum refining, consumer goods, building materials
La Coop fédérée (QC)	3,951	106	10,429	Pork and poultry, petroleum, farm supply
Agropur Co-opérative (QC)	3,345	3,459	5,441/435	Dairy products
United Farmers of Alberta Co-operative (AB)	1,747	228,166	723/435	Petroleum, farm supplies, building materials
Calgary Co-op Association (AB)	1,062	440,643	1,623/1,717	Supermarket, petroleum, travel agency, pharmacy
Co-op Atlantic (NB)	558	94	766/370	Wholesale, food, petroleum, farm supply
Red River Co-op (MB)	450	203,429	19	Petroleum retail
Gay Lea Foods Co-operative (ON)	442	3,485	498/46	Dairy products
Exceldor coopérative avicole (QC)	395	235	855	Meat processing and marketing, slaughtering
Hensall District Co-operative (ON)	338	4,486	299/79	Grain marketing, petroleum, farm supply

Source: Co-operatives Secretariat 2013.

APPENDIX 4

International Co-operative Principles

1. Voluntary and Open Membership

Co-operatives are voluntary organizations, open to all persons able to use their services and willing to accept the responsibilities of membership, without gender, social, racial, political, or religious discrimination.

2. Democratic Member Control

Co-operatives are democratic organizations controlled by their members, who actively participate in setting their policies and making decisions. Men and women serving as elected representatives are accountable to the membership. In primary co-operatives, members have equal voting rights (one member, one vote), and co-operatives at other levels are also organized in a democratic manner.

3. Member Economic Participation

Members contribute equitably to, and democratically control, the capital of their co-operative. At least part of that capital is usually the common property of the co-operative. Members usually receive limited compensation, if any, on capital subscribed as a condition of membership. Members allocate surpluses for any or all of the following purposes: developing their co-operative, possibly by setting up reserves, part of which at least would be indivisible; benefiting members in proportion to their transactions with the co-operative; and supporting other activities approved by the membership.

4. Autonomy and Independence

Co-operatives are autonomous, self-help organizations controlled by their members. If they enter into agreements with other organizations, including governments, or raise capital from external sources, they do so on terms that ensure democratic control by their members and maintain their co-operative autonomy.

5. Education, Training, and Information

Co-operatives provide education and training for their members, elected representatives, managers, and employees so they can contribute effectively to the development of their co-operatives. They inform the general public – particularly young people and opinion leaders – about the nature and benefits of co-operation.

6. Co-operation among Co-operatives

Co-operatives serve their members most effectively and strengthen the co-operative movement by working together through local, national, regional, and international structures.

7. Concern for Community

Co-operatives work for the sustainable development of their communities through policies approved by their members.

Source: ICA 2010.

APPENDIX 5

Major Proposed IPL Transmission Lines, 2014

Province	Project	Proponent	Timeline	IPL information
BC	Juan de Fuca Cable Project	Sea Breeze Power Corp.	Planning stages	Underwater 550 MW HVDC transmission line from Vancouver Island near Victoria to Washington State near Port Angeles
Saskatchewan and Alberta	Wind Spirit Project	Rocky Mountain Power and Grasslands Renewable Energy	Estimated completion by 2018	3,000 MW of nameplate capacity wind energy from four quadrants: Alberta, Montana, Saskatchewan, and North Dakota
Alberta	Montana-Alberta Tie-Line	Enbridge Power Inc.	Completed in 2013	230 kV, 300 MW, and 345-kilometre transmission line connecting southern Alberta and northern Montana
Alberta	Northern Lights/Alberta Electric System Operator's ten-year plan	TransCanada Alberta Electric System Operator	Early stages of development and planning	500 kV, 3,000 MW HVDC line from northern Oregon to Edmonton, with a possible extension to Fort McMurray

▾ **APPENDIX 5**

Province	Project	Proponent	Timeline	IPL information
Manitoba	Bipole III	Manitoba Hydro	Estimated completion by 2017	Modification of the existing 500 kV line from Dorsey Converter Station to Minnesota
Quebec	Northern Pass Transmission Project	Northeast Utilities	Estimated completion 2017	Quebec to New England, 330 km, 1,200 MW capacity project
Quebec	Champlain Hudson Power Express	Transmission developers, Hydro-Québec	Estimated completion 2017	539 km, 1,000 MW HVDC submarine line linking Montreal to Yonkers
Newfoundland and Labrador and Quebec	Lower Churchill development	Newfoundland and Labrador Hydro (with Emera)	Operational by 2017	Series of new lines linking Churchill Falls to Maine, including one 1,100 km, HVDC 900 MW Labrador Island Transmission Link, and another 180 km, 500 MW subsea line from Bottom Brook to Lingan
New Brunswick	Maritimes to northeastern United States	New Brunswick System Operator	Operational by 2017	1,200–1,500 MW capacity; HVDC IPL

Sources: Manitoba Hydro 2014; Nalcor Energy 2012; NEB 2009, 2010b, 2011.

APPENDIX 6

Ontario FIT and RESOP Prices

Date	Policy		Source	Rates (¢/kWh)	Project size	Community-targeted support
2006	RESOP	Guaranteed price (twenty-year contract) for new renewable generation under 10 MW	Wind Water Biomass Biogas Landfill gas	11	< 10 MW	n/a
			Solar	42		
2009	FIT/microFIT Program	Guaranteed price (twenty-year contract) for new renewable generation differentiated by source and size of project	Solar (rooftop)	80.2	≤ 10 kW	+1¢/kWh for community projects and +1.5¢/kWh for First Nations projects with more that 50% participation. Projects with a minimum 10% interest can earn a portion of the adder.
				71.3	> 10 kW ≤ 250 kW	
				63.5	> 250 kW ≤ 500 kW	
				53.9	> 500 kW	
			Solar (groundmount)	64.2	≤ 10 kW	
				44.3	> 10 kW	
			Wind	13.5	All	
			Water	13.1	≤ 10 MW	
				12.2	> 10 MW ≤ 50 MW	
			Biomass	13.8	≤ 10 MW	

APPENDIX 6

Date	Policy	Source	Rates (¢/kWh)	Project size	Community-targeted support
		Biogas (on farm)	13	> 10 MW	
			19.5	≤ 100 kW	
			18.5	> 100 kW ≤ 250 kW	
		Biogas	16	≤ 500 kW	
			14.7	> 500 kW ≤ 10 MW	
			10.4	> 10 MW	
		Landfill gas	11.1	≤ 10 MW	
			10.3	> 10 MW	
FIT 2.0 (2012)	FIT/microFIT Program. Guaranteed price (twenty-year contract) for new renewable generation differentiated by source and size of project	Solar (rooftop)	54.9	≤ 10 kW	Community and Aboriginal adders remain at 1 and 1.5¢/kWh respectively. The minimum ownership is raised from 10% to 15% to qualify for partial adder.
			54.8	> 10 kW ≤ 100 kW	
			53.9	> 100 kW ≤ 500 kW	
			48.7	> 500 kW	
		Solar (groundmount)	44.5	≤ 10 kW	
			44.3	> 10 kW	
			38.8	> 10 kW ≤ 500 kW	
			35	> 500 kW ≤ 5 MW	
			34.7	> 5 MW	
		Wind	11.5	All	
		Water	13.1	≤ 10 MW	
			12.2	> 10 MW ≤ 50 MW	
		Biomass	13.8	≤ 10 MW	
			13	> 10 MW	

Date	Policy	Source	Rates (¢/kWh)	Project size	Community-targeted support
		Biogas (on farm)	19.5	≤ 100 kW	
			18.5	> 100 kW ≤ 250 kW	
		Biogas	16	≤ 500 kW	
			14.7	> 500 kW ≤ 10 MW	
			10.4	> 10 MW	
		Landfill gas	11.1	≤ 10 MW	
			10.3	> 10 MW	
FIT 3.0 (prices as of January 2014)	FIT/microFIT Program	Guaranteed price (twenty-year contract) for new renewable generation differentiated by source and size of project, max. 500 kW			FIT 3.0 rules reduced the size to "small projects" as part of a directive from the Department of Energy. Adders lower for 15–50% share, higher for more than 50%. Aboriginal adder 1.5/0.75 Community participation 1.0/0.5 Municipal or public-sector entity 1.0/0.5
		Solar (rooftop)	39.6	≤ 10 kW	
			34.5	> 10 kW ≤ 100 kW	
			32.9	> 100 kW	
		Solar (groundmount)	29.1	≤ 10 kW	
			28.8	> 10 kW	
		Wind	11.5	All	
		Water	14.8	All	
		Renewable biomass	15.6	All	
		Biogas (on farm)	26.5	≤ 100 kW	
			21.0	> 100 kW ≤ 250 kW	
		Biogas	16.4	All	
		Landfill gas	7.7	All	

Notes

Chapter 1: A Climate for Change

1 There are, of course, many conceptualizations of what makes an organization, or a society, for that matter, democratic. A full exploration of these nuances of democratic theory is beyond the scope of this book, though several useful explorations of "deepening" democracy have informed this work: participatory (Fung and Wright 2003), discursive (Dryzek 1994), deliberative (Gutmann and Thompson 1996; Johnson 2008), and economic (Wood 1995).

2 Interviewees were contacted based on their employment in a provincial energy or economic development department, employment in a provincial co-operative association, or involvement in the development of community energy projects. Approximately one-third of those contacted agreed to be interviewed. Many of those who agreed to participate in the study elected to be anonymous. One hundred and forty potential interviewees were contacted on the basis of their association with co-operatives or with federal and provincial energy policy making.

3 Of course, not all states, or citizens within a state, are affected equally by these crises (Johnston, Gismondi, and Goodman 2006a).

4 The definition of the term "neoliberal," together with its conceptual utility, is contested. Boas and Gans-Morse (2009) point out how infrequently the term is defined, how normatively charged it is, and the breadth of diverse principles and practices it covers. Others, like Peck (2010) and Harvey (2005), have developed a sophisticated understanding of both the core principles of neoliberal theory and the diverse mechanisms of its implementation in practice.

5 During the 1940s and 1950s, rural electrification co-operatives developed in Alberta and Quebec to extend power lines to rural towns and farms.

6 Distributed generation is an electricity sector structure – also known as embedded generation or decentralized generation – wherein generation of power is dispersed through the power system, often with small facilities operating at the local level rather than concentrated in a few large generation facilities.

7 Although private utilities were often the first to provide electricity, it was the public sector that developed and extended much of the electricity system across the country (Froschauer 1999). Alberta was a notable exception, as much of the rural electrification there took place by way of co-operatives rather than a Crown corporation.

8 In many provinces, new renewables like wind, small hydro, and solar power form an increasing share of the generation mix. Chapter 5 outlines these developments by province but highlights the fact that ever-increasing demand, export pressures, and profit-drive growth are not necessarily leading to a significant enough shift toward a "green" energy future.

9 Co-operatives can be profit-making enterprises. Some American definitions of the social economy thus discount them (Salamon, Sokolowski, and Anheier 2000). Others, in the Canadian, European, and Latin American movements, tend to include co-operatives, since the profits are redistributed among local stakeholders (workers, consumers, residents, etc.).

10 The Columbia River Treaty arrangements present one important counterpoint to this argument.

11 Nuclear was rejected, as these shifts followed closely on the heels of the Chernobyl disaster.

12 This is also evident in Canada; Etcheverry illustrates how a collaborative network of activists, researchers, and policy makers led to the adoption of "paradigm-shifting" policy options, including the 2009 Green Energy and Green Economy Act in Ontario (Etcheverry 2013, 240).

13 For more on FIT designs see Chapters 5 and 6, as well as Gipe (2007, 2010), and REN21 (2014).

14 The largest co-operative wind-power generation projects – the 102 MW Bear Mountain Wind Park and the 45 MW Lamèque wind farm – are either partially or fully owned by private developers.

15 Multiple contracts can be awarded to the same organization, so contracted numbers should not be taken as indicative of the total number of co-ops. Indeed, establishing baseline data for the number of generating co-ops in operation is very difficult given the paucity of national data collection in the sector. Projects often go dormant, even after contracts are awarded. The most detailed data at this point is collected by the Federation of Community Power Co-ops in its annual sector survey.

Chapter 2: Governing Sustainability

1 Although Marx has been criticized in the past for championing industrial development (by Herman Daly and Robyn Eckersley, for example), work by James O'Connor (1998), Chris Williams (2010), Paul Burkett (2006), John Bellamy Foster (2002, 2009),

and Joel Kovel (2007) has shed new light on Marx's insights into alienation and exploitation of the natural environment.

Chapter 3: Co-operatives in Canadian Political Economy

1 Co-operatives are not the only organizations with a member-owned structure. Farmer-owned companies, nonprofits, and a range of other ownership structures are not co-operatives but share characteristics of co-operatives. Indeed, in the electricity sector in Germany, many of the German farmer organizations that developed wind projects were "community" (i.e., locally) based but not co-operatives per se.

2 According to the Government of Canada's *Status of Co-operatives in Canada* 2012 report, when credit unions are factored in, 17 million Canadians are members, or nearly half of Canada's population (Government of Canada 2012, 5).

3 This figure is based on a national survey by Industry Canada with a response rate of 65 percent of all registered co-operatives, so likely significantly understates the total membership (Industry Canada 2015, 3).

4 The role of investment for profit is contested in the co-operative movement. Attempts to loosen the rules on outside investment and the member-user link that provide a return on capital rather than patronage dividends are often seen as capitalistic and compromising co-operative principles (Alkalay personal interview, April 15, 2010).

5 Patronage works in proportion to what the co-op used. The more a co-op buys from Federated Co-operative, the more savings it gets in patronage; likewise, the more an individual shops at the co-op store locally, the more he or she receives in either cash or product. Because of Canada Revenue Agency holdbacks, members get 35 percent to 95 percent of their allocation. You need to buy a certain amount to qualify (this is so the co-op doesn't waste its time with 50-cent transactions). There are also age bylaws wherein members who reach sixty-five years of age can have their equity in the co-op paid back to them.

6 Most co-operatives are regulated under provincial acts. If a co-operative operates in more than two provinces (with fixed offices), it falls under the federal legislation (CCA 2011b). Without formal incorporation at the federal or provincial level, a co-operative is unable to legally use the term "co-op" or "co-operative" in its name.

Chapter 4: International Forces for Power-Sector Restructuring

1 In Canada, these include Plutonic Power (now Alterra), TransAlta, TransCanada, and Brookfield.

2 Electricity system capacity is different from *installed capacity,* which refers to the maximum electricity that a particular plant or source can produce, which can be further differentiated by *nameplate* capacity (maximum technically possible)

versus *actual* capacity (predicted output given local resources and other constraints). System administrators need to maintain a cushion of excess capacity to deal with temporary fluctuations into the grid.

3 Elements of electricity conform to a more narrow use of the term in economics as a specific good that is nonrivalrous and nonexcludable: for example, reliability, security of supply (Houldin 2005, 61), and voltage management on the grid.

4 Although electricity itself is not storable, some components (e.g., dams in hydro systems) provide storage capacity and as battery technology continues to improve these issues are likely to change. Still, what is technically possible still forms only one part of the transformational challenge. We still require funds, sites, and political will.

5 This is certainly the case both for public construction of major transmission lines for industrial use and for the construction of new transmission and distribution services for widespread distributed generation.

6 Hydroelectric power is bankable insofar as it can be stored in reservoirs and generation can be rapidly increased or decreased, as required by demand, by letting more water flow through the turbines.

Chapter 5: Continental, Private, and Green(er)?

1 Ontario and its commitment to phase out coal completely by 2014, as well as Nova Scotia's public mandate that its private utility add more renewable sources, are examples that exhibit a stronger environmental policy stance.

2 Pressure from FERC in the 1990s has led to functional separation of most provincial utilities exporting to the United States. Vertical integration remains through the common provincial ownership of the utility and its parts, but separate business units and organizations for transmission (like the BC Transmission Corporation or TransÉnergie) and generation and distribution aspects are created.

3 After Norway (95%), Brazil (80%), and Venezuela (68%) (IEA 2012b, 10).

4 Not all fossil fuels are created equal. Natural gas generation, for instance, has a significantly lower GHG intensity than coal or diesel generation.

5 Large-scale hydro also has negative environmental impacts, as every generation source does, which include flooding of often prime agricultural land, displacement of human and animal populations from large areas of land, and disruption of fish populations (Froschauer 1999). The benefits of hydro as a firm power source and the economic efficiencies that arise from a large-scale project may, in some cases, trump the alternatives in a life-cycle analysis, but a diversity of renewable sources suited to different human and natural geographic conditions is critically important.

6 The savings promised from this outsourcing in 2003 were meant to be in the order of $250 million over ten years (McMartin 2010).

7 Despite this growth, Canada still has one of the most underdeveloped wind resources in the world. Germany, which is twenty-eight times smaller in terms of land mass than Canada, has ten times more installed wind capacity (Valentine 2010).

Only 1.5 percent of Canada's total electricity generation in 2012 was from wind power.

Chapter 6: Electricity Co-operatives

1 Co-operatives also exist that generate small amounts of power for use in a building or small collection of buildings (Argenta Power Co-op in British Columbia is one). Tracing and cataloguing all of these falls outside the scope of this book. Generation co-operatives also include many co-operatives that act for many years as networking co-operatives while they work on developing a generation project.

2 In practice, categorizing the activities of co-operatives is complex, so I have assigned them to categories by their primary business area. There is often overlap between different types of co-operatives. For example, distribution co-operatives provide retail functions for their customers. Furthermore, educational co-operatives include those, like the former Community Power Fund, that provide financing and policy advocacy functions.

3 As of yet, there is no national survey of electricity co-operatives. The data for this table was compiled from a wide range of sources, including project websites, surveys by provincial associations, energy ministry reports where contracts have been awarded, and the historical co-operatives survey by the Co-operatives Secretariat, which was only updated to 2010. The numbers compiled likely significantly underestimate the number of consumer and networking co-operatives, as well as new renewable-electricity co-operatives, across the country. The best sector profiles at present are in the province of Ontario (see FCPC 2013, 2014, 2015).

4 Information on these newly incorporated energy co-operatives in Ontario is available from the Financial Services Commission of Ontario: https://www.fsco.gov.on.ca/. These new co-operatives have all been coded here as "generation" co-operatives, though many will also provide consumer and networking functions.

5 "Community power" is a term used to recognize a significant (usually more than 50 percent) ownership in a power project by, for example, co-operatives, First Nations, local landowners, and municipalities. Some of the controversies over this term are addressed in Chapter 8.

6 Franchise areas – granted to rural electrification associations in the United States and to gas co-operatives in Alberta – give the co-operative exclusive rights to service provision in a given geographic area.

7 Co-operatives during this early period also played a small role in (off-grid) power generation. The Argenta Water Power Co-op in British Columbia's Kootenay region, for example, was established in 1954 by a group of American Quakers to supply seventeen residences in their community.

8 Only about 40 percent of these were active, according to Doiron (2008).

9 Many co-operatives, such as SolarShare in Ontario with more than thirty solar projects in operation, have multiple contracts.

10 The community and First Nations adders are based on a "community participation level." The maximum adder for community and municipal participation (1 cent per kWh) is allocated to projects with 50 percent or more community ownership.

Projects with between 15 percent and 50 percent participation can receive a 0.5 cent adder under FIT 3.0 rules. The adder is 1.5 cents and 0.75 cents respectively for Aboriginal participation.

11 Prior to these modifications, co-operatives were required to do at least 50 percent of their business with members, and since renewable generation co-ops sell to the grid, they had difficulty incorporating until the changes passed.

12 The government of the Northwest Territories also has a fund to help community and Aboriginal groups build renewable generation: the Community Renewable Energy Fund. It provides up to half the project cost, up to $50,000 per year.

13 Quebec has, however, the most developed framework for co-operative development more generally in Canada (see Chapter 3).

14 Interestingly, in 2011, a new electricity policy was announced that reforms NB Power into an integrated utility, reversing the trend in other provinces (but similar to the reversal of the BC Transmission Corporation, in British Columbia).

Chapter 7: Off the Ground and on the Grid

1 The stalled/inactive section is significantly understated, as information on these (since they are inactive) is less readily available. It is very possible that many co-operatives never make it to the incorporation phase and fail to establish a web presence or register with local co-op associations, and so would be missed in the data collection for this project. Those listed as "stalled" here have either publicly announced that their project is on hold or have been unable to move forward for ten years from the date of incorporation.

2 The actual capacity is closer to 650 kW, since it had to be installed at a slightly lower height than manufacturer recommends because of zoning issues.

Chapter 8: Co-operative Networks and the Politics of Community Power

1 The Community Power Fund provided start-up for a range of co-operatives, including the Local Initiative for Future Energy (LIFE) Co-operative, TREC's OurPower, Windfall Ecology Centre (Pukwis project), Windy Hills Caledon, Barrie Windcatchers, and ZooShare.

Chapter 9: Empowering Electricity

1 Public electric utilities continue to be exempted under these arrangements, but private utilities and direct subsidies are not.

Glossary

---

capacity The amount of power a given electricity system can sustain. When referring to generation facilities, "nameplate capacity" refers to the maximum amount of power under ideal conditions that the technology can generate.

community power Electricity generation projects owned and controlled by First Nations, farmers, small businesses, nonprofits, co-operatives, and sometimes municipalities.

communitywashing Misleading claims about the local benefits of a product, policy, or activity made in order for it to gain greater public acceptance. *See also* greenwashing.

co-operative A flexible organizational form that is owned and managed by its membership. A co-operative can be structured as nonprofit or for-profit, and around various member groups – workers, consumers, producers – or various stakeholders.

co-operative difference The ways, both material and ideational, in which co-operatives are thought to be different from private, investor-owned businesses. These include democratic control through a one member, one vote structure; prioritization of member needs over profitability; voluntary and open membership; and the international statement of co-operative principles.

deep ecology Also known as deep or radical green. This perspective in the environmental literature rejects anthropocentrism. Deep greens advocate significant structural change (economic, social, technological, spiritual) in human societies and take issue with the current focus on economic growth and individual responsibility in promoting environmental sustainability.

double movement Concept developed by Karl Polanyi to describe the re-assertion of social control over market forces via protective legislation or restrictive associations.

eco-localism The view that environmental sustainability is best advanced by local self-reliant economic communities. Social economy advocates interested in environmental sustainability are often eco-locals.

electricity restructuring A process whereby integrated electricity systems, once thought to be natural monopolies, are broken up into generation, distribution, transmission, and retail components. Private actors play a larger role in restructured systems through either outright privatization of formerly public utility assets or dominating new areas, such as renewable power generation. Power markets (pools) are also sometimes created.

empowered participatory governance A model of governance attempting to deepen democracy put forward by Archon Fung and Erik Olin Wright. It advocates participatory socioeconomic processes of local ownership, deliberation, and empowerment in order to enhance both normative goals of distributive justice and positive ones of effective resilience and complex problem solving.

feedback loop In the literature on sustainability, a feedback loop refers to eco-social links (often geographically based) that produce informational flows needed for systems assessment, reassessment, and adaptations. Feedback loops are severed when systems of social organization and governance are stretched across increasingly long distances and decision makers are separated (in both space and time) from affected users.

feed-in tariff (FIT) A policy tool used to stimulate new renewable-electricity generation by providing long-term standard offer contracts with grid access, a fixed price based on the cost of generation plus a "reasonable" profit, and long-term stable purchase agreements. Tariff rates can be structured to differentiate between various technologies, project sizes, actors, and locales. FITs are an increasingly popular procurement mechanism internationally for the development of new renewable technologies and also for encouraging distributed generation.

governance The complex of actors and processes of power at multiple levels beneath, within, between, and above states, whether public or private. Governance structures authority, decision making, and accountability.

greenwashing Misleading claims of the environmental benefits of a product, policy, or activity, made in order for it to gain greater public acceptance.

independent power producer A nonutility generator of power that sells power to utilities for transmission to end users. A range of actors may own these facilities: investor-owned companies, co-operatives, municipalities, or First Nations.

neoliberalism The ideology of socioeconomic governance premised on a reliance on market-based resource allocation, the rollback of state expenditure on public services, privatization, and deregulation.

renewable electricity Electricity generated from renewable sources such as wind, solar, and hydro, as opposed to fossil fuel–based (nonrenewable) generation. Significant differences in how "renewable" is defined exist between jurisdictions.

set-aside A renewable generation policy tool that allocates a portion of either grid capacity or procurement of power to a specific actor or technology.

social ecology A perspective in green political theory that locates the roots of ecological problems in social and economic structures of hierarchy and capitalism. Social greens are critical of deep ecology's lack of attention to issues of justice and distribution. They are also critical of mainstream society's domination of nature and seek to find a balance between anthropocentrism and ecocentrism.

social economy A broad sector between the public and the private that incorporates co-operatives, credit unions, mutuals, and nonprofits. Social economy organizations often prioritize and incorporate social (and/or environmental) goals, as well as those of profit. Sometimes referred to as the "third sector."

social enterprise Businesses structured to generate income in order to meet social or environmental goals.

solidarity economy A part of the social economy wherein organizations are explicitly connected to promoting social and economic justice. The definition is still being developed but generally includes a specific concern for issues of labour and social justice.

sustainability Most commonly understood in environmental terms as the capacity to meet the needs of the present without compromising the needs of the future. The concept is generally understood to require the reconciliation of three pillars: environmental, economic, and social equity.

References

Abbott, Malcolm. 2001. "Is the Security of Electricity Supply a Public Good?" *Electricity Journal* 14 (7): 31–33. http://dx.doi.org/10.1016/S1040-6190(01)00224-X.

Achenbach, Joel. 2010. "The 21st Century Grid." *National Geographic* 218 (1): 122–38.

Adkin, Laurie. 2009. *Environmental Conflict and Democracy in Canada.* Vancouver: UBC Press.

Agence France-Presse. 2011. "Japan Challenges Canada at WTO over Green Energy Programme." June 18. http://www.ictsd.org/i/news/biores/109512/.

AGRIS Solar Co-operative. 2012. *Member Update October.* http://agrissolar.coop/news/Member%20Update%20-04October2012.pdf.

–. 2014. *Member Update December.* http://agrissolar.coop/2014/12/december-solar-gardens-update/

Akorede, Mudathir, Hashim Hizam, Ishak Aris, and Mohd-Zainal Kadir. 2010. "Re-emergence of Distributed Generation in Electric Power Systems." *Energy & Environment* 21 (2): 75–92. http://dx.doi.org/10.1260/0958-305X.21.2.75.

Albert, Michael. 2003. *Parecon: Life after Capitalism.* London: Verso.

Albo, Greg. 2006. "The Limits of Eco-localism: Scale, Strategy, Socialism." In *Coming to Terms with Nature,* edited by Colin Leys and Leo Panitch, 337–63. Halifax: Fernwood.

Amin, Ash, ed. 2009. *The Social Economy: International Perspectives on Economic Solidarity.* New York: Zed Books.

Amin, Ash, Cameron Angus, and Ray Hudson. 2002. *Placing the Social Economy.* London: Routledge.

Anderson, John A. 2009. "Electricity Restructuring: A Review of Efforts around the World." *Electricity Journal* 22 (3): 70–86. http://dx.doi.org/10.1016/j.tej.2009.02.017.

Andersson, Krister, and Elinor Ostrom. 2008. "Analyzing Decentralized Resource Regimes from a Polycentric Perspective." *Policy Sciences* 41 (1): 71–93. http://dx. doi.org/10.1007/s11077-007-9055-6.

Asiskovitch, Sharon. 2011. "Dismantling the Welfare State from the Left? Localization of Global Ideas in the Case of Israel's 1998 Public Housing Law." *Global Social Policy* 11 (1): 69–87. http://dx.doi.org/10.1177/1468018110391997.

Bäckstrand, Karin. 2010. *Environmental Politics and Deliberative Democracy: Examining the Promise of New Modes of Governance.* Cheltenham, UK: Edward Elgar. http://dx.doi.org/10.4337/9781849806411.

Baker, Susan. 2014. "Governance." In *Critical Environmental Politics,* edited by Carl Death, 100–10. New York: Routledge.

Bakker, Karen. 2010. "The Limits of 'Neoliberal Natures': Debating Green Neoliberalism." *Progress in Human Geography* 34 (6): 715–35. http://dx.doi. org/10.1177/0309132510376849.

Barclay, Richard A. 2009. *Feed-in Tariffs.* Ann Arbor: Michigan Electric Cooperative Association.

Barkin, J. Samuel. 2006. "Discounting the Discount Rate: Ecocentrism and Environmental Economics." *Global Environmental Politics* 6 (4): 56–72. http://dx.doi. org/10.1162/glep.2006.6.4.56.

Barry, John. 2012. *The Politics of Actually Existing Unsustainability: Human Flourishing in a Climate-Changed Carbon Constrained World.* New York: Oxford University Press. http://dx.doi.org/10.1093/acprof:oso/9780199695393.001.0001.

Barry, John, Geraint Ellis, and Clive Robinson. 2008. "Cool Rationalities and Hot Air: A Rhetorical Approach to Understanding Debates on Renewable Energy." *Global Environmental Politics* 8 (2): 67–98. http://dx.doi.org/10.1162/ glep.2008.8.2.67.

BC Hydro. 2000. *Annual Report.* Vancouver: BC Hydro.

–. 2001. *Annual Report.* Vancouver: BC Hydro.

–. 2011. *Independent Power Producers Currently Supplying Power to BC Hydro.* Vancouver: BC Hydro.

–. 2013. *Annual Report.* Vancouver: BC Hydro.

–. 2014. *Annual Report.* Vancouver: BC Hydro.

Beder, Sharon. 2003. *Power Play: The Fight to Control the World's Electricity.* New York: New Press.

Bell, Jeff, and Tim Weis. 2009. *Greening the Grid: Powering Alberta's Future with Renewable Energy.* Calgary: Pembina Institute.

Bergen, William E. 1984. *Co-operation: It's Good for Canada.* Saskatoon: Co-Enerco.

Blakes Lawyers. 2008. *Overview of Electricity Regulation in Canada.* Toronto: Blake, Cassels, and Graydon.

Blay-Palmer, Alison. 2011. "Sustainable Communities: An Introduction." *Local Environment: The International Journal of Justice and Sustainability* 16 (8): 747–52. http://dx.doi.org/10.1080/13549839.2011.613235.

Blue, Ian. 2009. "Off the Grid: Federal Jurisdiction and the Canadian Electricity Sector." *Dalhousie Law Journal* 32 (2): 339–66.

Boas, Taylor, and Jordan Gans-Morse. 2009. "Neoliberalism: From New Liberal Philosophy to Anti-Liberal Slogan." *Studies in Comparative International Development* 44 (2): 137–61. http://dx.doi.org/10.1007/s12116-009-9040-5.

Bolinger, Mark A. 2005. "Making European-Style Community Wind Power Development Work in the US." *Renewable & Sustainable Energy Reviews* 9 (6): 556–75.
http://dx.doi.org/10.1016/j.rser.2004.04.002.

Bookchin, Murray. 1999. "Social Ecology versus Deep Ecology: A Challenge for the
Ecology Movement." In *Philosophical Dialogues: Arne Næss and the Progress of
Ecophilosophy*, edited by Nina Witoszek and Andrew Brennan, 281–301. New
York: Rowman and Littlefield.

Bouffard, François, and Daniel Kirschen. 2008. "Centralised and Distributed Electricity Systems." *Energy Policy* 36 (12): 4504–8. http://dx.doi.org/10.1016/j.
enpol.2008.09.060.

Bowles, Samuel, and Herbert Gintis. 1986. *Democracy and Capitalism: Property,
Community, and the Contradictions of Modern Social Thought.* New York: Basic
Books.

Boyer, Robert, and Yves Saillard, eds. 2002. *Régulation Theory: The State of the Art.*
New York: Routledge.

Bradford, Neil. 2005. *Place-Based Public Policy: Towards a New Urban and Community Agenda for Canada.* Research report F51. Ottawa: Canadian Policy Research
Networks.

Bratt, Duane. 2006. *The Politics of CANDU Exports.* Toronto: University of Toronto
Press.

–. 2012. *Canada and the Global Nuclear Revival.* Montreal: McGill-Queen's University Press.

Brecher, Jeremy, Tim Costello, and Brendan Smith. 2008. "Labor's War on
Global Warming." *The Nation,* March 10. http://www.thenation.com/article/
labors-war-global-warming.

British Columbia. 2011. *Review of BC Hydro.* Victoria: Government of British Columbia.

–. 2013. *10 Year Plan Means Predictable Rates.* November 26. Victoria: Ministry of
Energy and Mines.

Brooks, David. 2006. "Power to the Public." *Alternatives* 32 (1): 37–38.

Brundtland, Gro Harlem. 1987. *Our Common Future: The World Commission on
Environment and Development.* Oxford: Oxford University Press.

Bulkeley, Harriet. 2005. "Reconfiguring Environmental Governance: Towards a Politics of Scales and Networks." *Political Geography* 24 (8): 875–902. http://dx.doi.
org/10.1016/j.polgeo.2005.07.002.

Burkett, Paul. 2006. *Marxism and Ecological Economics: Toward a Red and Green
Political Economy.* Leiden, Netherlands: Koninklijke Brill NV.

Burtraw, Dallas, Karen Palmer, and Marin Heintzelman. 2000. *Electricity Restructuring: Consequences and Opportunities for the Environment.* Washington, DC:
Resources for the Future.

Byrne, John, Noah Toly, and Leigh Glover, eds. 2006. *Transforming Power: Energy,
Environment and Society in Conflict.* Piscataway, NJ: Transaction.

Calvert, John. 2007. *Liquid Gold: Energy Privatization in British Columbia.* Black
Point, NS: Fernwood.

Campion-Smith, Bruce. 2011. "700 Environment Canada Jobs on the Chopping Block."
Toronto Star. August 3. http://www.thestar.com/news/canada/2011/08/03/700_
environment_canada_jobs_on_the_chopping_block.html.

Canada. 2006. "Communities under Pressure: The Role of Co-operatives and the Social Economy." In *Co-operative Membership and Globalization: Creating Social Cohesion through Market Relations.* Ottawa: Policy Research Initiative, Government of Canada.

Canadian Electricity Association. 2010. "Canada's Electricity Industry: Background and Challenges." PowerPoint presentation. http://www.electricity.ca/media/Electricity101/Electricity101.pdf.

CanWEA (Canadian Wind Energy Association). 2009. *Canada Milestone as Wind Energy Now in Every Province.* December 30. http://canwea.ca/media/release/release_e.php?newsId=70.

CAREA (Central Alberta Rural Electrification Association). 2010. *CAREA at a Glance.* Inisfail, AB: CAREA.

—. 2011. *CAREA at a Glance.* Inisfail, AB: CAREA.

—. 2012. *CAREA and Other REAs Joint Press Release.* February 10. http://centralalbertarea.blogspot.ca/2012/02/carea-and-other-alberta-reas-issue.html.

Carter, Neil. 1996. "Worker Co-operatives and Green Political Theory." In *Democracy and Green Political Thought,* edited by Brian Doherty and Marius de Geus, 55–77. New York: Routledge. http://dx.doi.org/10.4324/9780203449554_chapter_3.

Catney, Philip, Andrew Dobson, Sarah Marie Hall, Sarah Hards, Sherilyn MacGregor, Zoe Robinson, Mark Ormerod, and Simon Ross. 2013. "Community Knowledge Networks: An Action-Orientated Approach to Energy Research." *Local Environment* 18 (4): 506–20. http://dx.doi.org/10.1080/13549839.2012.748729.

Catney, Philip, Sherilyn MacGregor, Andrew Dobson, Sarah Marie Hall, Sarah Royston, Zoe Robinson, Mark Ormerod, and Simon Ross. 2013. "Big Society, Little Justice? Community Renewable Energy and the Politics of Localism." *International Journal of Justice and Sustainability*: 1–16. doi: 10.1080/13549839.2013.792044.

Cavanagh, John, and Jerry Mander. 2005. *Alternatives to Economic Globalization: A Better World Is Possible.* San Francisco: Berrett-Koehler Press.

CCA (Canadian Co-operative Association). 2011a. *Co-operatives Helping Fuel a Green Economy.* Ottawa: CCA.

—. 2011b. "Incorporating Your Co-op." July 20. http://www.coopscanada.coop/.

CCPA (Canadian Centre for Policy Alternatives), Parkland Institute, and Polaris Institute. 2006. *Fuelling Fortress America: A Report on the Athabasca Tar Sands and U.S. Demands for Canada's Energy.* Ottawa: CCPA, Parkland Institute, and Polaris Institute.

Christianson, Russ. 2011. "Danish Wind Co-ops Can Show Us the Way." WindWorks. November 16. http://www.wind-works.org/cms/.

Clarkson, Stephen. 2002. *Uncle Sam and Us: Globalization, Neoconservatism and the Canadian State.* Toronto: University of Toronto Press.

Clément, Michel, and Caroline Bouchard. 2008. "Taux de survie des coopératives au Québec." Quebec City: Government of Quebec.

Climate Action Network. 2011. *Canada Wins Fossil of the Year Award in Durban.* http://climateactionnetwork.ca/2011/12/09/canada-wins-fossil-of-the-year-award-in-durban/.

Cohen, Marjorie Griffin. 2004. "International Forces Driving Electricity Deregulation in the Semi-periphery: The Case of Canada." In *Governing Under Stress:*

Middle Powers and the Challenge of Globalization, edited by Marjorie Griffin Cohen and Stephen Clarkson, 175–94. London: Zed.

–. 2006a. *Why Canada Needs a National Energy Plan.* Vancouver: Canadian Centre for Policy Alternatives.

–. 2006b. "Electricity Restructuring's Dirty Secret." In *Nature's Revenge: Reclaiming Sustainability in an Age of Corporate Globalization,* edited by Josée Johnston, Michael Gismondi, and James Goodman, 73–95. Peterborough, ON: Broadview Press.

–. 2007. "Imperialist Regulation: U.S. Electricity Market Designs and Their Problems for Canada and Mexico." In *Whose Canada? Continental Integration, Fortress North America and the Corporate Agenda,* edited by Ricardo Grinspun, Yasmine Shamsie, and Maude Barlow, 439–58. Montreal: McGill-Queen's University Press.

Community Power Fund. 2010. http://www.cpfund.ca.

Co-operatives Secretariat. 2010a. *Overview of Co-operatives in Canada, 2007.* Ottawa: Government of Canada.

–. 2010b. *About Co-ops in Canada.* Ottawa: Government of Canada.

–. 2011. *Top 50 Non-Financial Co-operatives in Canada 2009.* Ottawa: Government of Canada.

–. 2013. *Top 50 Non-Financial Co-operatives in Canada 2010.* Ottawa: Government of Canada.

Côté, Daniel. 2000. "Co-operatives in the New Millennium: The Emergence of a New Paradigm." In *Canadian Co-operatives in the Year 2000: Memory, Mutual Aid and the Millennium,* edited by Nora Russell, 250–66 Saskatoon: Centre for the Study of Co-operatives.

Cottier, Thomas, Garba Malumfashi, Sfya Metteotti-Berkutova, and Olga Nartova. 2010. *Energy in WTO Law and Policy.* Bern: World Trade Institute.

Cox, Robert. 1996. *Approaches to World Order.* Cambridge Studies in International Relations. Cambridge: Cambridge University Press. http://dx.doi.org/10.1017/CBO9780511607905.

Cumbers, Andrew. 2012. *Reclaiming Public Ownership: Making Space for Economic Democracy.* London: Zed.

–. 2013. "Making Space for Economic Democracy: The Danish Wind Power Revolution." Paper presented at the UNRISD conference "Potential and Limits of Social and Solidarity Economy," May 6–8, Geneva, Switzerland.

Curl, John. 2010. "The Cooperative Movement in Century 21." *Affinities* 4 (1): 12–29.

Curry, Bill, and Shawn McCarthy. 2011. "Canada Formally Abandons Kyoto Protocol on Climate Change." *Globe and Mail.* December 12. http://www.theglobeandmail.com/news/politics/canada-formally-abandons-kyoto-protocol-on-climate-change/article2268432/.

Curtis, Fred. 2003. "Eco-localism and Sustainability." *Ecological Economics* 46 (1): 83–102. http://dx.doi.org/10.1016/S0921-8009(03)00102-2.

Cuthill, Michael. 2010. "Strengthening the 'Social' in Sustainable Development: Developing a Conceptual Framework for Social Sustainability in a Rapid Urban Growth Region in Australia." *Sustainable Development* 18 (6): 362–73. http://dx.doi.org/10.1002/sd.397.

Daly, Herman. 1996. *Beyond Growth*. Boston: Beacon Press.

Daly, Herman, and John Cobb. 1989. *For the Common Good: Redirecting the Economy toward Community, the Environment and a Sustainable Future*. Boston: Beacon Press.

Danish Energy Agency. 2013. "Key Figures." http://www.ens.dk/en/info/facts-figures/key-figures/danish-key-figures.

Danish Energy Association. 2009. *Danish Electricity Supply 2008*. Frederiksberg C, Denmark: Dansk Energi.

Datamonitor. 2010. *Electricity Industry Profile: Canada*. New York: Datamonitor PLC.

Davies, Anna Ray. 2009. "Does Sustainability Count? Environmental Policy, Sustainable Development and the Governance of Grassroots Sustainability Enterprise in Ireland." *Sustainable Development* 17 (3): 174–82. http://dx.doi.org/10.1002/sd.374.

de Peuter, Greig, and Nick Dyer-Withford. 2010. "Commons and Co-operatives." *Affinities* 1: 30–56.

de Villemeur, Etienne Billette, and Pierre-Olivier Pineau. 2010. "Environmentally Damaging Electricity Trade." *Energy Policy* 38 (3): 1548–58. http://dx.doi.org/10.1016/j.enpol.2009.11.038.

De Young, Raymond, and Thomas Princen, eds. 2012. *The Localization Reader: Adapting to the Coming Downshift*. Cambridge, MA: MIT Press.

Debor, Sarah. 2014. *The Socio-Economic Power of Renewable Energy Production Cooperatives in Germany*. Wuppertal, Germany: Wuppertal Institute.

Della Porta, Donatella, and Sidney Tarrow, eds. 2004. *Transnational Protest and Global Activism*. New York: Rowman and Littlefield.

Deller, Steven, Ann Hoyt, and Brent Hueth. 2009. *Economic Impact of Co-operatives*. Madison: University of Wisconsin Center for Cooperatives.

Devine-Wright, Patrick, ed. 2011. *Renewable Energy and the Public: From NIMBY to Participation*. London: Earthscan.

Dobson, Andrew. 2005. "Ecological Citizenship." In *Debating the Earth: The Environmental Politics Reader*, edited by John Dryzek and David Schlosberg, 596–607. London: Oxford University Press.

Doern, Bruce, and Monica Gattinger. 2003. *Power Switch: Energy Regulatory Governance in the Twenty-First Century*. Toronto: University of Toronto Press.

Doern, Bruce, and Glen Toner. 1985. *The Politics of Energy: The Development and Implementation of the Nep*. Agincourt, ON: Methuen Publications.

Doiron, Marie-Josee. 2008. "Les cooperatives et l'electrification rurale du Québec, 1945–1964." PhD diss., Université du Québec à Trois-Rivières.

Dolphin, Frank, and John Dolphin. 1993. *Country Power: The Electrical Revolution in Rural Alberta*. Edmonton: Plains Publishing.

Driesen, David, ed. 2009. *Economic Thought and US Climate Change Policy*. Cambridge, MA: MIT Press.

Dryzek, John S. 1992. "Ecology and Discursive Democracy: Beyond Liberal Capitalism and the Administrative State." *Capitalism, Nature, Socialism* 3 (2): 18–42. http://dx.doi.org/10.1080/10455759209358485.

–. 1994. *Discursive Democracy: Politics, Policy and Political Science*. New York: Cambridge University Press.

–. 2002. *Deliberative Democracy and Beyond: Liberals, Critics, Contestations.* Oxford: Oxford University Press. http://dx.doi.org/10.1093/019925043X.001.0001.

Dubash, Navroz, and James Williams. 2006. "The Political Economy of Electricity Liberalization." In *Transforming Power: Energy, Environment and Society in Conflict,* edited by John Byrne, Noah Toly, and Leigh Glover, 155–88. Piscataway, NJ: Transaction.

Duguid, Fiona. 2007. "Part of the Solution: Developing Sustainable Energy through Co-operatives and Learning." PhD diss., University of Toronto.

Dupré, Ruth, Michel Patry, and Patrick Joly. 1996. *The Politics and Regulation of Hydroelectricity: The Case of Québec in the Thirties.* CIRANO Scientific Series 96s-02.

Durant, Darrin. 2009. "The Trouble with Nuclear." In *Nuclear Waste Management in Canada: Critical Issues, Critical Perspectives,* edited by Darrin Durant and Genevieve Fuji Johnson, 11–30. Vancouver: UBC Press.

Durant, Darrin, and Genevieve Fuji Johnson, eds. 2009. *Nuclear Waste Management in Canada: Critical Issues, Critical Perspectives.* Vancouver: UBC Press.

Eckersley, Robin. 2004. *The Green State: Rethinking Democracy and Sovereignty.* London: MIT Press.

Economist. 2011. "(Government) Workers of the World Unite!" January 6. http://www.economist.com/node/17849199.

Egan, Tim, and Eli Turk. 2008. *Providing Reliable Energy in a Time of Constraints: A North American Concern.* Toronto: Canadian Electricity Association.

Ehrlich, Paul. 1968. *The Population Bomb.* New York: Ballantine.

Emond, Katie. 2010. *PowerWedges: Wind and Cogeneration Opportunities for Alberta Thought Leaders Forum Report.* Calgary: Calgary Economic Development and Pembina Institute.

ENVINT Consulting and Ontario Sustainable Energy Association. 2008. *Guide to Developing a Community Renewable Energy Project in North America.* Toronto: Commission for Environmental Cooperation.

Environment Canada. 2013. *National Inventory Report 1990–2011: Greenhouse Gas Sources and Sinks in Canada.* Ottawa: Government of Canada.

Etcheverry, José. 2013. "From Green Energy to Smart Growth: Practical Lessons from the Renewable Energy Movement for Agricultural Land Protection and Sustainability Activists." In *Health and Sustainability in the Canadian Food System,* edited by Rod McRae and Elisabeth Abergel, 223–42. Vancouver: UBC Press.

Evans, Simon. 2014. "How the IPCC Is Sharpening Its Language on Climate Change." The Carbon Blog 2014 (cited September 3). http://www.carbonbrief.org/blog/2014/09/how-the-ipcc-is-sharpening-its-language-on-climate-change/.

Faber, Daniel. 2008. *Capitalizing on Environmental Injustice: The Polluter-Industrial Complex in the Age of Globalization.* Lanham, MD: Rowman and Littlefield.

Fairbairn, Brett. 1990. "Co-operatives as Politics: Membership, Citizenship, and Democracy." In *Co-operative Organizations and Canadian Society,* edited by Murray Fulton, 129–40. Toronto: University of Toronto Press.

–. 2003. *Living the Dream: Membership and Marketing in the Co-operative Retailing System.* Saskatoon: Centre for the Study of Co-operatives.

Fairbairn, Brett, June Bold, Murray Fulton, Lou Hammond Ketilson, and Daniel Ish. 1995. *Co-operatives and Community Development: Economics in Social Perspective.* Saskatoon: Centre for the Study of Co-operatives.

Fairbairn, Brett, and Nora Russell, eds. 2004. *Co-operative Membership and Globalization.* Saskatoon: Centre for the Study of Co-operatives.

–, eds. 2014. *Co-operative Canada: Empowering Communities and Sustainable Businesses.* Vancouver: UBC Press.

Faucher, Albert. 1947. "Co-operative Trends in Canada." *Annals of the American Academy of Political and Social Science* 253 (1): 184–89. http://dx.doi.org/10.1177/000271624725300126.

FCL (Federated Co-operatives Limited). 2011. *2010 Annual Report.* Saskatoon: FCL.

FCPC (Federation of Community Power Co-operatives). 2013. *Community Power White Paper.* Toronto: Federation of Community Power Co-operatives.

–. 2014. *Annual Report FY 2013.* Toronto: FCPC.

–. 2015. *Annual Report FY 2014.* Toronto: FCPC.

Fitzpatrick, Tony. 2002. "Green Democracy and Ecosocial Welfare." In *Environment and Welfare,* edited by Tony Fitzpatrick and Michael Cahill, 61–81. London: Palgrave.

–, ed. 2014. *International Handbook on Social Policy and the Environment.* London: Edward Elgar. http://dx.doi.org/10.4337/9780857936134.

Folke, Carl. 2006. "Resilience: The Emergence of a Perspective for Social-Ecological Systems Analyses." *Global Environmental Change* 16 (3): 253–67. http://dx.doi.org/10.1016/j.gloenvcha.2006.04.002.

Fontan, Jean-Marc, and Eric Shragge, eds. 2000. *Social Economy: International Debates and Perspectives.* Montreal: Black Rose Books.

Forsyth, Tim. 2010. "Panacea or Paradox? Cross-Sector Partnerships, Climate Change, and Development." *Wiley Interdisciplinary Reviews: Climate Change* 1 (5): 683–96. http://dx.doi.org/10.1002/wcc.68.

Foster, John Bellamy. 2002. *Ecology against Capitalism.* New York: Monthly Review Press.

–. 2009. *The Ecological Revolution: Making Peace with the Planet.* New York: Monthly Review Press.

Freire, Paulo. 2000. *Pedagogy of Freedom: Ethics, Democracy and Civic Courage.* New York: Rowman and Littlefield.

Freitas, Walmir, Eduardo N. Asada, Ahmed F. Zobaa, and James S. McConnach. 2007. "Policy and Economic Issues of Electrical Power and Energy Systems." *International Journal of Global Energy Issues* 27 (3): 253–61. http://dx.doi.org/10.1504/IJGEI.2007.014347.

Froschauer, Karl. 1999. *White Gold: Hydroelectric Development in Canada.* Vancouver: UBC Press.

Fulton, Murray, ed. 1990. *Co-operative Organizations and Canadian Society: Popular Institutions and the Dilemmas of Change.* Toronto: University of Toronto Press.

–. 2001. *New Generation Co-operative Development in Canada.* Saskatoon: Centre for the Study of Co-operatives.

Fulton, Murray, and Lou Hammond Ketilson. 1992. "The Role of Cooperatives in Communities: Examples from Saskatchewan." *Journal of Agricultural Cooperation* 7: 15–42.

Fung, Archon, and Erik Olin Wright. 2003. *Deepening Democracy: Institutional Innovations in Empowered Participatory Governance*. Real Utopias Project. London: Verso.

Galuzzo, Teresa Welsh. 2005. *Small Packages, Big Benefits: Economic Advantages of Local Wind Projects*. Mount Vernon, IA: Iowa Policy Project.

Gamble, Andrew, Steve Ludlam, Andrew Taylor, and Stephen Wood, eds. 2007. *Labour, the State, Social Movements and the Challenge of Neo-liberal Globalisation*. Manchester: Manchester University Press.

Gattinger, Monica. 2010. "Canada's Energy Policy Relations in North America." In *Borders and Bridges: Canada's Policy Relations in North America*, edited by Monica Gattinger and Geoffrey Hale, 139–47. Don Mills, ON: Oxford University Press.

Gattinger, Monica, and Geoffrey Hale. 2010. *Borders and Bridges: Canada's Policy Relations in North America*. Don Mills, ON: Oxford University Press.

Genalta Power. 2011. *2011 Oil Sands Co-Generation Report*. Calgary: Oil Sands Developers Group.

Gertler, Michael. 2001. *Rural Co-operatives and Sustainable Development*. Saskatoon: Centre for the Study of Co-operatives.

Gillis, Justin. 2014. "Sun and Wind Alter Landscape, Leaving Utilities Behind." *New York Times*. September 13.

Gipe, Paul. 2007. *Wind Energy Cooperative Development in Anglophone Canada*. Report for Canadian Cooperative Association. Wind-Works. http://www.wind-works.org/cms/.

–. 2010. "Provincial Feed-in Tariffs Spurring Community Power." Wind-Works. http://www.wind-works.org/cms/.

–. 2013. "Two Steps Forward, One Back: Ontario Cancels Feed-in Tariffs for Large Projects." *Renewable Energy World*. June 10.

–. n.d. *Renewables without Limits*. Toronto: Ontario Sustainable Energy Association. http://www.ontario-sea.org/Storage/22/1375_RenewablesWithoutLimits.pdf.

Girvitz, Geoff, and Judith Lipp. 2005. "Community Power Canadian Style: Wind-Share and the Toronto Renewable Energy Co-operative." *Refocus* 6 (1): 28–30. http://dx.doi.org/10.1016/S1471-0846(05)00290-8.

Godoy, Julio. 2011. "Fukushima Forces Europe to Rethink Nuclear Energy." *Global Issues* (Washington, DC). April 1.

Golob, Ursa, Klement Podnar, and Marko Lah. 2009. "Social Economy and Social Responsibility: Alternatives to Global Anarchy of Neoliberalism?" *International Journal of Social Economics* 36 (5): 626–40. http://dx.doi.org/10.1108/03068290910954068.

Goodman, Roger J. 2010. *Power Connections: Canadian Electricity Trade and Foreign Policy*. Toronto: Canadian International Council.

Gossen, L.E. 1975. *An Introduction to Cooperatives*. Saskatoon: Co-operative College of Canada.

Government of Canada. 2012. *Status of Co-operatives in Canada*. Report of the Special Committee on Co-operatives. Ottawa: Library of Parliament. http://www.

parl.gc.ca/content/hoc/Committee/411/COOP/Reports/RP5706528/cooprp01/cooprp01-e.pdf.

Government of Manitoba. n.d. Co-operative Development Tax Credit for Co-operatives and Credit Unions. Winnipeg. https://www.gov.mb.ca/finance/business/ccredits.html#co-opdtc.

Government of Ontario. 2009. *Green Energy and Economy Act – Bill 150*. http://www.ontario.ca/laws/statute/s09012.

Graefe, Peter. 2006. "Social Economy Policies as Flanking for Neoliberalism: Transnational Policy Solutions, Emergent Contradictions, Local Alternatives." *Politics & Society* 23 (3): 69–86.

Gramsci, Antonio. 1971. *Selections from the Prison Notebooks*. London: International Publishers.

Gratwick, Katharine N., and Anton Eberhard. 2008. "Demise of the Standard Model for Power Sector Reform and the Emergence of Hybrid Power Markets." *Energy Policy* 36 (10): 3948–60. http://dx.doi.org/10.1016/j.enpol.2008.07.021.

Green Energy Act Alliance. 2011. *Ontario Feed-in Tariff Review, Joint Submission*. Toronto: Green Energy Act Alliance and Shine Ontario.

Guimond, Pierre. 2010. *Canada's Electricity Industry: Background and Challenges*. Ottawa: Canadian Electricity Association.

Gutmann, Amy, and Denis Thompson. 1996. *Democracy and Disagreement*. Cambridge, MA: Belknap Press.

Hahnel, Robin. 2007. "Eco-localism: A Constructive Critique." *Capitalism, Nature, Socialism* 18 (2): 62–78. http://dx.doi.org/10.1080/10455750701366444.

Halifax Media Co-op. 2010. *Nova Scotia's Neo-Liberal Energy Policies*. Halifax: Halifax Media Co-op. http://halifax.mediacoop.ca/blog/macdonaldtomw/4407.

Hall, David. 1999. *Electricity Restructuring, Privatisation and Liberalisation: Some International Experiences*. London: Public Services International.

Hampton, Howard. 2003. *Public Power*. Toronto: Insomniac Press.

Hand, Maureen M., Samuel F. Baldwin, Ed DeMeo, John M. Reilly, Trieu Mai, Doug Arent, Gian Porro, Mike Meshek, and Debra Sandor. 2012. *Renewable Electricity Futures Study*. Golden, CO: National Renewable Energy Laboratory.

Harcourt, Mark, and Geoffrey Wood, eds. 2004. *Trade Unions and Democracy: Strategies and Perspectives*. Manchester: Manchester University Press.

Harden-Donahue, Andrea, and Andrea Peart. 2009. *Green, Decent and Public*. Toronto: Canadian Labour Congress and Council of Canadians.

Hardin, Garrett. 1968. "The Tragedy of the Commons." *Science* 162 (3859): 1243–48. http://dx.doi.org/10.1126/science.162.3859.1243.

Harrison, Kathryn. 2012. "Multilevel Governance and American Influence on Canadian Climate Policy: The California Effect vs. the Washington Effect." *Zeitschrift für Kanada-Studien* 32 (2): 45–64.

Harvey, David. 2005. *A Brief History of Neoliberalism*. Oxford: Oxford University Press.

–. 2010. *The Enigma of Capital*. London: Profile Books.

Heiman, Michael K. 2006. "Expectations for Renewable Energy under Market Restructuring: The U.S. Experience." *Energy & Environment* 31 (6): 1052–66.

Heiman, Michael K., and Barry D. Solomon. 2004. "Power to the People: Electric Utility Restructuring and the Commitment to Renewable Energy." *Annals of the Association of American Geographers* 94 (1): 94–116. http://dx.doi.org/10.1111/j. 1467-8306.2004.09401006.x.

Held, David, and Anthony McGrew. 2002. *Governing Globalization: Power, Authority and Global Governance*. Cambridge, UK: Polity Press.

Herman, Edward S., and Noam Chomsky. 2002. *Manufacturing Consent*. New York: Pantheon Books.

Heynen, Nik, James McCarthy, Scott Prudham, and Paul Robbins. 2007. *Neoliberal Environments: False Promises and Unnatural Consequences*. New York: Routledge.

Hoberg, George. 2011. "Playing Defence: Early Responses to Conflict Expansion in the Oil Sands Policy Subsystem." *Canadian Journal of Political Science* 44 (3): 507–27. http://dx.doi.org/10.1017/S0008423911000473.

Hoberg, George, and Ian H. Rowlands. 2012. "Green Energy Politics in Canada: Comparing Electricity Policies in BC and Ontario." Paper presented at American Political Science Association annual meeting, New Orleans, August 30 to September 2.

Hoffman, Steven M., and Angela High-Pippert. 2009. "Community Energy: A Social Architecture for an Alternative Energy Future." *Bulletin of Science, Technology & Society* 25 (5): 387–401.

Holburn, Guy L.F. 2012. "Assessing and Managing Regulatory Risk in Renewable Energy: Contrasts between Canada and the United States." *Energy Policy* 45: 654–65. http://dx.doi.org/10.1016/j.enpol.2012.03.017.

Homer-Dixon, Thomas. 2007. *The Upside of Down: Catastrophe, Creativity and the Renewal of Civilization*. Toronto: Knopf Canada.

–. 2009. *Carbon Shift: How the Twin Crises of Oil Depletion and Climate Change Will Define the Future*. New York: Random House.

Horlick, Gary, and Christiane Schuchhardt. 2002. *NAFTA Provisions and the Electricity Sector*. Montreal: Commission for Environmental Cooperation of North America.

Houldin, Russell W. 2005. "Lost Economies of Integration and the Costs of Creating Markets in Electricity Restructuring: Evidence from Ontario." *Electricity Journal* 18 (8): 45–54. http://dx.doi.org/10.1016/j.tej.2005.08.001.

Howe, Bruce, and Frank Klassen. 1996. *The Case of B.C. Hydro: A Blueprint for Privatization*. Vancouver: Fraser Institute.

Howlett, Karen, Bertrand Marotte, and Richard Blackwell. 2012. "WTO Rules against Ontario in Green Energy Dispute." *Globe and Mail*. November 20.

Howlett, Michael, M. Ramesh, and Anthony Perl. 2003. *Studying Public Policy, Policy Cycles and Policy Subsystems*. 2nd ed. Don Mills, ON: Oxford University Press.

Howse, Robert, and Gerald Heckman. 1996. "The Regulation of Trade in Electricity: A Canadian Perspective." In *Ontario Hydro at the Millennium*, edited by Ronald J. Daniels, 103–55. Montreal: McGill-Queen's University Press.

Huckle, John. 2012. "Even More Sense and Sustainability." *Environmental Education Research* 18 (6): 845–58. http://dx.doi.org/10.1080/13504622.2012.665851.

Hunt, Tim. 2010. "Workers of the World, Cooperate!" *Energy Bulletin*. June 1. http://www.energybulletin.net/stories/2010-06-29/workers-world-cooperate.

Hydro-Québec. 2003. *Comparing Power Generation Options*. Quebec City: Hydro-Québec.

–. 2010. *Liste des Soumissions Retenues*. Quebec City: Hydro-Québec.

–. 2011. "History of Electricity in Québec." http://www.hydroquebec.com/history -electricity-in-quebec/.

–. 2014. "Wind Power." http://www.hydroquebec.com/sustainable-development/energy -environment/wind-power.html.

–. 2015. "Comparison of Electricity Prices in Major North American Cities." Quebec City: Hydro-Québec. http://www.hydroquebec.com/publications/en/docs/ comparaison-electricity-prices/comp_2015_en.pdf.

ICA (International Co-operative Alliance). 2010. "Statement on the Co-operative Identity." http://ica.coop/coop/principles.html.

–. 2015. "Facts and Figures." http://ica.coop/en/whats-co-op/co-operative-facts -figures.

IEA (International Energy Agency). 2005. *Lessons from Liberalised Electricity Markets*. Paris: Organisation for Economic Co-operation and Development.

–. 2012a. *Denmark 2011 Review*. Paris: Organisation for Economic Co-operation and Development.

–. 2012b. *Technology Roadmap: Hydropower*. Paris: OECD/IEA.

ILO (International Labour Organization). 2013. *Providing Clean Energy and Energy Access through Cooperatives*. Geneva: ILO.

Industry Canada. 2015. *Co-operatives in Canada in 2010*. Ottawa: Government of Canada.

IPCC (Intergovernmental Panel on Climate Change). 2007. *Climate Change 2007*. Fourth assessment report. Geneva: IPCC.

–. 2011. *Special Report on Renewable Energy Sources and Climate Change Mitigation*. New York: Cambridge University Press.

–. 2014. *Climate Change 2014: Mitigation of Climate Change*. Geneva: IPCC.

Ipsos Reid. 2010. *Canadian Co-operative Association Baseline Awareness Survey*. Ottawa: Canadian Co-operative Association and Ipsos Reid.

Jaccard, Mark, and Jeffrey Simpson. 2007. *Hot Air: Meeting Canada's Climate Change Challenges*. Toronto: McClelland and Stewart.

Jacobsson, S., and V. Lauber. 2006. "The Politics and Policy of Energy System Transformation – Explaining the German Diffusion of Renewable Energy Technology." *Energy Policy* 34 (3): 256–76. http://dx.doi.org/10.1016/j.enpol.2004.08.029.

Jenkins, Nick, Janaka Ekanayake, and Goran Strbac. 2009. *Distributed Generation: The Institution of Engineering and Technology*. London: IET.

Jensen, Pia C., and Steen Hartvig Jacobsen. 2009. *Wind Turbines in Denmark*. Copenhagen: Danish Energy Agency.

Jessop, Bob. 1995. "The Regulation Approach, Governance, and Post-Fordism: Alternative Perspectives on Economic and Political Change." *Economy and Society* 24 (3): 307–33. http://dx.doi.org/10.1080/03085149500000013.

–. 2002. "Liberalism, Neoliberalism, and Urban Governance: A State-Theoretical Perspective." *Antipode* 34 (3): 452–72. http://dx.doi.org/10.1111/1467-8330.00250.

Johnson, Genevieve Fuji. 2004. "Ethical Policy in an Age of Risk, Uncertainty, and Futurity." PhD diss., University of Toronto.

–. 2008. *Deliberative Democracy for the Future: The Case of Nuclear Waste Management in Canada*. Toronto: University of Toronto Press.

–. 2009. "Deliberative Democratic Practices in Canada: An Analysis of Institutional Empowerment in Three Cases." *Canadian Journal of Political Science* 42 (3): 679–703. http://dx.doi.org/10.1017/S0008423909990072.

–. 2011. "The Limits of Deliberative Democracy and Empowerment: Elite Motivation in Three Canadian Cases." *Canadian Journal of Political Science* 44 (1): 137–59. http://dx.doi.org/10.1017/S0008423910001058.

Johnston, Josée, Michael Gismondi, and James Goodman. 2006a. "Politicizing Exhaustion: Eco-social Crisis and the Geographic Challenge for Cosmopolitans." In *Nature's Revenge: Reclaiming Sustainability in an Age of Corporate Globalization,* edited by Josée Johnston, Michael Gismondi, and James Goodman, 13–35. Peterborough, ON: Broadview Press.

–, eds. 2006b. *Nature's Revenge: Reclaiming Sustainability in an Age of Corporate Globalization.* Peterborough, ON: Broadview Press.

Jordan, Bill. 2010. *Why the Third Way Failed.* London: Policy Press.

Joskow, Paul L. 2009. "US vs. EU Electricity Reforms Achievement." In *Electricity Reform in Europe: Towards a Single Energy Market,* edited by Jean-Michel Glachant and François Lévêque, xiii–xxix. London: Edward Elgar.

Kalmi, Panu. 2007. "The Disappearance of Co-operatives from Economic Textbooks." *Cambridge Journal of Economics* 31 (4): 625–47. http://dx.doi.org/10.1093/cje/bem005.

Kasmir, Sharryn. 1996. *The Myth of Mondragón: Cooperatives, Politics, and Working-Class Life in a Basque Town.* New York: SUNY Press.

Keevers, Lynn, Chris Skykes, and Lesley Treleaven. 2008. "Partnership and Participation: Contradictions and Tensions in the Social Policy Space." *Australian Journal of Social Issues* 43 (3): 459–77.

Kerr, Tom. 2008. *IEA CHP/DHC Country Scorecard: Denmark.* Paris: International Energy Agency.

Ketilson, Lou Hammond, Michael Gertler, Murray Fulton, Roy Dobson, and Leslie Polsom. 1998. *The Social and Economic Importance of the Co-operative Sector in Saskatchewan.* Research report prepared for Saskatchewan Department of Economic and Co-operative Development. Saskatoon: Centre for the Study of Co-operatives.

Kovel, Joel. 2007. *The Enemy of Nature: The End of Capitalism or the End of the World?* Halifax: Fernwood.

Krupa, Joel, Lindsay Galbraith, and Sarah Burch. 2013. "Participatory and Multi-Level Governance: Applications to Aboriginal Renewable Energy Projects." *Local Environment* (ahead-of-print): 1–21.

Kumhof, Michael, and Romain Ranciere. 2010. *Inequality, Leverage and Crises.* IMF Working Paper WP/10/268. Washington, DC: International Monetary Fund. http://www.imf.org/external/pubs/ft/wp/2010/wp10268.pdf.

Kuntze, Jan-Christoph, and Tom Moerenhout. 2013. *Local Content Requirements and the Renewable Energy Industry – A Good Match?* Geneva: International Centre for Trade and Sustainable Development. http://dx.doi.org/10.7215/GP_IP_20130603.

Laidlaw, Alexander Fraser. 1980. *Co-operatives in the Year 2000.* Geneva: International Co-operative Alliance.

Lambert, Rob. 2007. "Self-Regulating Markets, Restructuring and the New Labour Internationalism." In *Labour, the State, Social Movements and the Challenge of Neo-Liberal Globalisation,* edited by Andrew Gamble, Steve Ludlam, Andrew Taylor, and Stephen Wood, 147–70. New York: Palgrave.

Larner, Wendy, and David Craig. 2005. "After Neo-Liberalism? Community Activism and Local Partnerships in Aotearoa, New Zealand." *Antipode* 37 (3): 402–24. http://dx.doi.org/10.1111/j.0066-4812.2005.00504.x.

Larsen, Ib. 2005. *Renewable Energy in District Heating in Denmark.* Copenhagen: Danish Energy Authority.

Lauersen, Birger. 2008. *Denmark – Answer to a Burning Platform: Chp/Dhc.* Paris: International Energy Agency.

Laville, Jean-Louis, Benoit Levesque, and Marguerite Mendell. 2007. "The Social Economy: Diverse Approaches and Practices in Europe and Canada." In *The Social Economy: Building Inclusive Economies,* edited by Antonella Noya and Emma Clarence, 155–87. Paris: Organisation for Economic Co-operation and Development. http://dx.doi.org/10.1787/9789264039889-7-en.

Laycock, David. 1990. "Democracy and Co-operative Practice." In *Co-operative Organizations and Canadian Society,* edited by Murray Fulton, 77–92. Toronto: University of Toronto Press.

LeBlanc, Alfred. 2006. "The Government of Canada and the Social Economy." *Horizons Policy Research Initiatives* 8 (2): 4–8.

Leblanc, Leo, and Denyse Guy. 2014. *CMC Annual Report.* Co-operatives and Mutuals Canada. http://canada.coop/sites/canada.coop/files/files/CMC_AR_E_Final_Web.pdf.

Lehtonen, Markku. 2004. "The Environmental-Social Interface of Sustainable Development: Capabilities, Social Capital, Institutions." *Ecological Economics* 49 (2): 199–214. http://dx.doi.org/10.1016/j.ecolecon.2004.03.019.

Lionais, Doug, and Harvey Johnstone. 2010. "Building the Social Economy Using the Innovative Potential of Place." In *Living Economics: Canadian Perspectives on the Social Economy, Co-operatives, and Community Economic Development,* edited by John J. McMurtry, 105–28. Toronto: Emond Montgomery.

Lipp, Judith. 2008. *Achieving Local Benefits: Policy Options for Community Energy in Nova Scotia.* Halifax: Nova Scotia Department of Energy.

Lipp, Judith, Émanuèlle Lapierre-Fortin, and J.J. McMurtry. 2012. *Renewable Energy Co-op Review: Scan of Models and Regulatory Issues.* Toronto: Measuring the Co-operative Difference Research Network. http://www.cooperativedifference.coop/featured-research/renewable-energy/.

Lloyd, Peter. 2007. "The Social Economy in the New Political Economic Context." In *The Social Economy: Building Inclusive Economies,* edited by Antonella Noya and Emma Clarence, 61–90. Paris: Organisation for Economic Co-operation and Development. http://dx.doi.org/10.1787/9789264039889-4-en.

Lord, Andrew. 2011 "Ontario Domestic Content Rules Facing WTO and NAFTA Challenges." https://www.dlapiper.com/en/canada/.

Loring, Joyce McLaren. 2007. "Wind Energy Planning in England, Wales and Denmark: Factors Influencing Project Success." *Energy Policy* 35: 2684–60.

Lovins, Amory. 1977. *Soft Energy Paths: Toward a Durable Peace.* San Francisco: Friends of the Earth.

Loxley, John, ed. 2007. *Transforming or Reforming Capitalism? Towards a Theory of Community Economic Development.* Black Point, NS: Fernwood.

Luke, Timothy. 2002. "Deep Ecology: Living as if Nature Mattered: Devall and Sessions on Defending the Earth." *Organization & Environment* 15 (2): 178–86. http://dx.doi.org/10.1177/10826602015002005.

Lyster, Rosemary. 2005. "The Implications of Electricity Restructuring for a Sustainable Energy Framework: What's Law Got to Do with It?" In *The Law of Energy for Sustainable Development,* edited by Adrian J. Bradbrook, Rosemary Lyster, Richard L. Ottinger, and Wang Xi, 415–48. New York: Cambridge University Press. http://dx.doi.org/10.1017/CBO9780511511387.028.

MacArthur, Julie. 2014. "Sustainability and the Social Economy in Canada: From Resource Reliance to Resilience?" In *International Handbook on Social Policy and the Environment,* edited by Tony Fitzpatrick, 274–99. London: Edward Elgar.

–. 2015. "Challenging Public Engagement: Participation, Deliberation and Power in Energy Policy Design and Implementation." *Journal of Environmental Studies and Sciences* (Online First, September 22): 1–10.

MacDonald, Douglas. 2007. *Business and Environmental Politics in Canada.* Peterborough, ON: Broadview Press.

MacGillivray, Anne, and Daniel Ish. 1992. *Co-operatives in Principle and Practice.* Centre for the Study of Co-operatives Occasional Paper no. 92 (01).

Macpherson, Crawford B. 1973. "Elegant Tombstones: A Note on Friedman's Freedom." In *Democratic Theory: Essays in Retrieval,* 143–56. Oxford: Clarendon.

–. 1977. *The Life and Times of Liberal Democracy.* Oxford: Oxford University Press.

MacPherson, Ian. 2008. "The Co-operative Movement and Social Economy Traditions: Reflections on the Mingling of Broad Visions." *Annals of Public and Cooperative Economics* 79 (3–4): 625–42. http://dx.doi.org/10.1111/j.1467-8292.2008.00373.x.

–. 2009. *A Century of Co-operation.* Ottawa: Canadian Co-operative Association.

Manczyk, Henry, and Michael D. Leach. n.d. *Combined Heat and Power Generation and District Heating in Denmark: History, Goals and Technology.* University of Rochester (cited January 2, 2010). http://blog.cleanenergy.org/files/2009/12/denmark.pdf.

Manitoba Hydro. 2014. "Bipole III Project." August. https://www.hydro.mb.ca/proj ects/bipoleIII/index.shtml.

Martin, Oliver. 2012. "The Dream That Failed." *Economist.* March 10.

Marvin, Simon, and Simon Guy. 1997. "Creating Myths Rather Than Sustainability: The Transition Fallacies of the New Localism." *International Journal of Justice and Sustainability* 2 (3): 311–18. doi: 10.1080/13549839708725536.

Marx, Karl. [1845] 1974. "Theses on Feuerbach." In Karl Marx and Frederick Engels, *The German Ideology,* 121–23. London: Lawrence and Wishart.

Mayo, Ed. 2011. *Co-operatives UK Urges Caution Over Open Public Services White Paper.* July 30. http://www.uk.coop/newsroom/co-operatives-uk-urges-caution -over-open-public-services-white-paper.

McBride, Stephen. 2005. *Paradigm Shift: Globalization and the Canadian State.* Halifax: Fernwood.

McLeod-Kilmurray, Heather, and Gavin Smith. 2010. "Unsustainable Development in Canada: Environmental Assessment, Cost-Benefit Analysis, and Environmental Justice in the Tar Sands." *Journal of Environmental Law and Practice* 21: 65–105.

McMartin, Will. 2010. "BC Liberals Owe Us a $65 Million Apology." *The Tyee*. May 3. http://thetyee.ca/Opinion/2010/05/03/LiberalsOweApology/.

McMurtry, J.J. 2004. "Social Economy as Political Practice." *International Journal of Social Economics* 31 (9): 868–78. http://dx.doi.org/10.1108/03068290410550656.

–, ed. 2010. *Living Economics: Canadian Perspectives on the Social Economy, Cooperatives, and Community Economic Development*. Toronto: Emond Montgomery.

Meyer, Neils. 2007. "Learning from Wind Energy Policy in the EU: Lessons from Denmark, Sweden and Spain." *European Environment* 17 (5): 347–62. http://dx.doi.org/10.1002/eet.463.

Meyer, Susan. 2009. "The Power of Numbers." *Peace Country Sun*. http://www.peacecountrysun.com/2009/12/23/the-power-of-numbers.

Ministre du Développement économique, de l'Innovation et de l'Exportation. 2011. *Cooperatives*. http://www.mdeie.gouv.qc.ca/objectifs/informer/cooperatives/.

Mitchell, Catherine. 2008. *The Political Economy of Sustainable Energy*. London: Palgrave Macmillan.

Möller, Bernd. 2010. "Spatial Analyses of Emerging and Fading Wind Energy Landscapes in Denmark." *Land Use Policy* 27 (2): 233–41. http://dx.doi.org/10.1016/j.landusepol.2009.06.001.

Moore, Oliver. 2010. "New Brunswick Voters Kick Liberals Out Over Resource Control." *Globe and Mail*. September 27. http://www.theglobeandmail.com/news/politics/new-brunswick-voters-kick-liberals-out-over-resource-control/article1729225/.

Munck, Ronald. 2002. "Globalization and Democracy: A New 'Great Transformation'?" *Annals of the American Academy of Political and Social Science* 581 (1): 10–21. http://dx.doi.org/10.1177/0002716202058001003.

Murphy, Raymond, and Maya Murphy. 2012. "The Tragedy of the Atmospheric Commons: Discounting Future Costs and Risks in Pursuit of Immediate Fossil-Fuel Benefits." *Canadian Review of Sociology* 49 (3): 247–70. http://dx.doi.org/10.1111/j.1755-618X.2012.01294.x.

Musall, Fabian David, and Onno Kuik. 2011. "Local Acceptance of Renewable Energy – A Case Study from Southeast Germany." *Energy Policy* 39 (6): 3252–60. http://dx.doi.org/10.1016/j.enpol.2011.03.017.

Næss, Arne. 1973. "The Shallow and the Deep, Long-Range Ecology Movement." *Inquiry* 16 (1–4): 95–100. http://dx.doi.org/10.1080/00201747308601682.

Nalcor Energy. 2012. "Lower Churchill Project." http://www.nalcorenergy.com/lower-churchill-project.asp.

Natural Resources Canada. 2008. "Overview of Canada's Energy Policy." http://www.nrcan.gc.ca/.

Neamtan, Nancy. 2002. "The Social and Solidarity Economy: Towards an 'Alternative' Globalization." Paper presented at "Citizenship and Globalization: Exploring Participation and Democracy in a Global Context" conference, Vancouver.

Neamtan, Nancy, and Rupert Downing. 2005. *Social Economy and Community Economic Development in Canada*. Montreal: Chantier de L'économie Sociale.

NEB (National Energy Board). 2000. *Exports and Imports of Electricity*. Calgary: NEB.
–. 2008. *Exports and Imports of Electricity*. Calgary: NEB.
–. 2009. *Canadian Energy Overview 2008*. Calgary: NEB.
–. 2010a. *Canadian Energy Overview 2009*. Calgary: NEB.
–. 2010b. *Canada's Energy Future: Infrastructure Challenges and Changes to 2020*. Calgary: NEB.
–. 2011. *North American Energy Market, Presentation to New England–Canada Business Council*. Calgary: NEB.
–. 2012. *Exports and Imports of Electricity*. Calgary: NEB.
–. 2013. *Exports and Imports of Electricity – Export Summary Report December 2013*. Calgary: NEB.
–. 2014. *Exports and Imports of Electricity – Export Summary Report December 2014*. Calgary: NEB.
Nelson, Sharon L. 1997. "Competition in Electricity: Transition from Almost a Public Good to Almost a Commercial Commodity." Keynote speech given at Washington Utilities and Transportation Commission, fourth annual Energy Resources, Washington: Conservation and Recycling Conference, April 22.
Netherton, Alexander. 2007. "The Political Economy of Canadian Hydro-Electricity: Between Old 'Provincial Hydros' and Neoliberal Regional Energy Regimes." *Canadian Political Science Review* 1 (1): 107–24.
New Brunswick. 2010. *The Community Energy Policy*. Fredericton: Department of Energy.
–. n.d. *Community Energy Policy*. Fredericton: Department of Energy and Mines. http://www2.gnb.ca/content/gnb/en/departments/energy/renewable/content/CommunityRenewableEnergy.html.
Newell, Peter. 2000. *Climate for Change: Non-State Actors and the Global Politics of the Greenhouse*. Cambridge: Cambridge University Press. http://dx.doi.org/10.1017/CBO9780511529436.
Newig, Jens, and Oliver Fritsch. 2009. "Environmental Governance: Participatory, Multi-Level – and Effective?" *Environmental Policy and Governance* 19 (3): 197–214. http://dx.doi.org/10.1002/eet.509.
Nikiforuk, Andrew. 2011. "Wikileaks Shines Light on Alberta's $16-Billion Electricity Scandal." *The Tyee*. May 26. http://thetyee.ca/News/2011/05/26/WikileaksAlbertaElectricity/.
Nishimura, Kensuke. 2012. "Grassroots Action for Renewable Energy: How Did Ontario Succeed in the Implementation of a Feed-in Tariff System?" *Energy, Sustainability and Society* 2 (1): 1–11. http://dx.doi.org/10.1186/2192-0567-2-6.
Nova Scotia Department of Energy. 2010. *Renewable Electricity Plan*. Halifax: Government of Nova Scotia.
–. 2014. *Report on the Review of the Community Feed-in Tariff Program*. Halifax: Government of Nova Scotia.
Nova Scotia Government. 2011. *Feed in Tariffs*. Halifax: Department of Energy.
–. 2013. "Community Renewable Energy Projects Come Online." News release. March 19. http://novascotia.ca/news/release/?id=20130319001.
–. 2014. "Community Feed-in Tariff Results Available." March 6. http://novascotia.ca/news/release/?id=20140306001.

–. 2015a. "Minister Announces COMFIT Review Results, End to Program." News release. http://novascotia.ca/news/release/?id=20150806001.

–. 2015b. "COMFIT." http://energy.novascotia.ca/renewables/programs-and-projects/comfit.

O'Connor, James. 1998. "Capitalism, Nature, Socialism: A Theoretical Introduction." *Capitalism Nature Socialism* 1: 11–39.

O'Connor, Martin, ed. 1994. *Is Capitalism Sustainable? Political Economy and the Politics of Ecology*. New York: Guilford Press.

OECD (Organisation for Economic Co-operation and Development). 2001. *Competition in Electricity Markets*. Paris: OECD.

–. 2004. *Economic Survey of Canada 2004*. Paris: OECD.

–. 2015. *Climate Change Mitigation: Policies and Progress*. Paris: OECD. http://www.oecd.org/environment/climate-change-mitigation-9789264238787-en.htm.

Ontario. 2011. *Ontario's Long-Term Energy Plan*. Toronto: Ministry of Energy.

–. 2012. *Ontario's Feed-in Tariff Program Two Year Review Report*. Toronto: Ministry of Energy.

–. 2013a. *Ontario Getting Out of Coal-Fired Generation*. January 9. http://news.ontario.ca/mei/en/2013/1/ontario-getting-out-of-coal-fired-generation.html.

–. 2013b. *Ontario Working with Communities to Secure Clean Energy Future*. May 30. http://news.ontario.ca/mei/en/2013/05/ontario-working-with-communities-to-secure-clean-energy-future.

OOAG (Ontario Office of the Auditor General). 2015. *Annual Report 2015: Energy Chapter Section 3.05*. http://www.auditor.on.ca/en/reports_en/en15/3.05en15.pdf.

OPA (Ontario Power Authority). 2009. "New Green Energy Projects Generate More Green Jobs." News release. Toronto: OPA.

–. 2010. *Bi-weekly FIT and microFIT Report Oct 12*. Toronto: OPA.

–. 2011. *Bi-weekly FIT and microFIT Report Dec 23*. Toronto: OPA.

–. 2012. *Bi-weekly FIT and microFIT Report Aug 7*. Toronto: OPA.

–. 2013a. *Draft Documents Fit 3.0*. Toronto: OPA.

–. 2013b. *Fit 2.1 Contract Offers*. August 20. http://fit.powerauthority.on.ca/sites/default/files/page/FIT_2.1_Application_List_Contracts_v5.1.pdf.

–. 2014. *Fit 3.0 Applications Summary Contract Offers*. July 29. http://fit.powerauthority.on.ca/sites/default/files/version3/FIT%203.0%20Application%20Summary_Contract%20Offers_v1_20140729.pdf.

Ophuls, William. 1973. "Leviathan or Oblivion?" In *Toward a Steady State Economy*, edited by Herman Daly, 215–30. San Francisco: Freeman.

Orr, Fay. 1989. *Harvesting the Flame: The History of Alberta's Rural Natural Gas Cooperatives*. Edmonton: Reidmore Books.

OSEA (Ontario Sustainable Energy Association). 2009. *A Green Energy Act for Ontario: Executive Summary*. Toronto: OSEA.

–. 2011. "WWEA Highlights the Importance of Community Power and Publishes Definition." Toronto: OSEA. http://www.ontario-sea.org/Page.asp?PageID=924&ContentID=3081.

Ostrom, Elinor. 1990. *Governing the Commons: The Evolution of Institutions for Collective Action*. Cambridge: Cambridge University Press. http://dx.doi.org/10.1017/CBO9780511807763.

–. 2002. "Policy Analysis in the Future of Good Societies." *Good Society* 11 (1): 42–48. http://dx.doi.org/10.1353/gso.2002.0013.

Paehlke, Robert C. 2008. *Some Like It Cold: The Politics of Climate Change.* Toronto: Between the Lines.

Pahl, Greg. 2007. "A Case Study in Community Wind: Denmark." In *The Citizen-Powered Energy Handbook: Community Solutions to a Global Crisis,* 71–77. White River, VT: Chelsea Green.

Panitch, Leo. 2007. *Renewing Socialism: Transforming Democracy, Strategy and Imagination.* London: Merlin Press.

Panitch, Leo, and Colin Leys, eds. 2006. *Coming to Terms with Nature: Socialist Register, 2007.* Halifax: Fernwood.

Patel, Raj. 2009. *The Value of Nothing: How to Reshape Market Society and Redefine Democracy.* London: Portobello.

Pateman, Carole. 1988. *Participation and Democratic Theory.* Cambridge: Cambridge University Press.

Peck, Jamie. 2010. *Constructions of Neoliberal Reason.* Oxford: Oxford University Press. http://dx.doi.org/10.1093/acprof:oso/9780199580576.001.0001.

Peck, Jamie, and Adam Tickell. 2002. "Neoliberalizing Space." *Antipode* 34 (3): 380–404. http://dx.doi.org/10.1111/1467-8330.00247.

Penner, Dylan. 2011. "Poll Suggests Harper Government Out of Step with Canadians." http://www.commondreams.org/newswire/2010/11/18/poll-suggests-harper-government-out-step-canadianshttp://canadians.org/media/energy/2010/18-Nov-10.html.

Peris, Jordi, Miriam Acebillo-Baque, and Carola Calabuig. 2011. "Scrutinizing the Link between Participatory Governance and Urban Environmental Management: The Experience in Arequipa during 2003–2006." *Habitat International* 35 (1): 84–92. http://dx.doi.org/10.1016/j.habitatint.2010.04.003.

Perkins, Rudy. 1998. "Electricity Deregulation, Environmental Externalities and the Limitations of Price." *Boston College Law Review* 39 (4): 993–1059.

Pilon, Dennis. 2001. *Canada's Democratic Deficit: Is Proportional Representation the Answer?* Toronto: CSJ Foundation for Research and Education.

Pirnia, Mehrdad, Jatin Nathwani, and David Fuller. 2011. "Ontario Feed-in-Tariffs: System Planning Implications and Impacts on Social Welfare." *Electricity Journal* 24 (8): 18–28. http://dx.doi.org/10.1016/j.tej.2011.09.009.

Polanyi, Karl. 1944. *The Great Transformation: The Political and Economic Origins of Our Time.* Boston: Beacon Press.

Princen, Thomas. 2001. "Consumption and Its Externalities: Where Economy Meets Ecology." *Global Environmental Politics* 1 (3): 11–30. http://dx.doi.org/10.1162/152638001316881386.

Princen, Thomas, Michael Maniates, and Ken Conca, eds. 2002. *Confronting Consumption.* Cambridge, MA: MIT Press.

Purcell, Mark. 2008. *Recapturing Democracy: Neoliberalization and the Struggle for Alternative Urban Futures.* New York: Routledge.

Purcell, Mark, and J. Christopher Brown. 2005. "Against the Local Trap: Scale and the Study of Environment and Development." *Progress in Development Studies* 5 (4): 279–97. http://dx.doi.org/10.1191/1464993405ps122oa.

Quarter, Jack. 1992. *Canada's Social Economy: Co-operatives, Non-profits and Other Community Enterprises*. Toronto: James Lorimer.

Québec. 2011. *Wind Energy Projects in Québec*. Report by the Ministère des Ressources Naturelles et de la Faune. Quebec City: Government of Quebec.

–. 2015. *A New Energy Policy for Québec: Renewable Energies*. Ministére de l'Énergie et des Ressources naturelles. Quebec City: Government of Quebec. http://www. politiqueenergetique.gouv.qc.ca/wp-content/uploads/Document4-Renewable_ energy.pdf.

Quezada, Víctor H. Méndez, Abbad Juan Rivier, and Tomás Gómez San Román. 2006. "Assessment of Energy Distribution Losses for Increasing Penetration of Distributed Generation." *IEEE Transactions on Power Systems* 21 (2): 533–40. http://dx.doi.org/10.1109/TPWRS.2006.873115.

Rees, William. 2010. "What's Blocking Sustainability? Human Nature, Cognition, and Denial." *Sustainability: Science, Practice, & Policy* 6 (2): 13–25.

REN21. 2009. *Renewables Global Status Report: 2009 Update*. Paris: REN21 Secretariat.

–. 2014. *Global Status Report*. Paris: REN21 Secretariat.

Restakis, John. 2010. *Humanizing the Economy: Co-operatives in the Age of Capital*. Gabriola Island, BC: New Society.

Restakis, John, and Evert A. Lindquist. 2001. *The Co-op Alternative: Civil Society and the Future of Public Services*. Toronto: Institute of Public Administration of Canada.

Rifkin, Jeremy. 2002. *The Hydrogen Economy*. New York: J.P. Tarcher/Putnam.

Robinson, John. 2007. "Clearing the Air on Climate Change." *Literary Review of Canada* 15 (8): 14–15.

Rosenau, James. 1995. "Governance in the Twenty-First Century." *Global Governance* 1 (13): 13–43.

–. 2003. *Distant Proximities: Dynamics beyond Globalization*. Princeton, NJ: Princeton University Press.

Rotmans, J., R. Kemp, and M. van Asselt. 2001. "More Evolution than Revolution: Transition Management in Public Policy." *Foresight: The Journal of Futures Studies, Strategic Thinking and Policy* 3 (1): 15–31. http://dx.doi.org/10.1108/ 14636680110803003.

Rowlands, Ian H. 2007. "The Development of Renewable Electricity Policy in the Province of Ontario: The Influence of Ideas and Timing." *Review of Policy Research* 24 (3): 185–207. http://dx.doi.org/10.1111/j.1541-1338.2007.00277.x.

Saint-Pierre, Jacques. 1997. *Histoire de la Coopérative fédérée de Québec: L'industrie de la terre*. Montreal: Institut Québecois de Recherche sur la Culture.

Salamon, Lester, Wojciech Sokolowski, and Helmut K. Anheier. 2000. *Social Origins of Civil Society: An Overview*. Baltimore: Johns Hopkins University Centre for Civil Society Studies.

Sandberg, L. Anders, and Tor Sandsberg. 2010. *The Chilly Climates of the Global Environmental Dilemma*. Ottawa: Canadian Centre for Policy Alternatives.

Sathaye, Jayant, Oswaldo Lucon, and Atiq Rahman. 2011. *IPCC Special Report on Renewable Energy Sources and Climate Change Mitigation: Renewable Energy in the Context of Sustainable Development*. Cambridge: Cambridge University Press.

Saunders, J. Owen. 2001. "North American Deregulation of Electricity: Sharing Regulatory Sovereignty." *Texas International Law Journal* 36: 167–73.

Scheer, Herman. 2007. *Energy Autonomy*. London: Earthscan.

Scholte, Jan Aart. 2003. *Democratizing the Global Economy: The Role of Civil Society*. Coventry, UK: University of Warwick Centre for the Study of Globalisation and Regionalisation.

Schugurensky, Daniel, and Erica McCollum. 2010. "Notes in the Margins: The Social Economy in Economics and Business Textbooks." In *Researching the Social Economy*, edited by Laurie Mook, Jack Quarter, and Sherida Ryan, 154–75. Toronto: University of Toronto Press.

Schumacher, Ernst F. 1973. *Small Is Beautiful: Economics as if People Mattered*. London: Blond and Briggs.

Sen, Amartya. 1999. *Development as Freedom*. Oxford: Oxford University Press.

Seyfang, Gill, and Adrian Smith. 2007. "Grassroots Innovations for Sustainable Development: Towards a New Research and Policy Agenda." *Environmental Politics* 16 (4): 584–603. http://dx.doi.org/10.1080/09644010701419121.

Shiva, Vandana. 1993. *Monocultures of the Mind: Perspectives on Biodiversity and Biotechnology*. London: Zed Books.

Shragge, Eric. 1997. *Community Economic Development: In Search of Empowerment and Alternatives*, edited by Eric Shragge. Montreal: Black Rose Books.

–. 2003. *Activism and Social Change: Lessons for Community and Local Organizing*. Toronto: University of Toronto Press.

Sinoski, Killy. 2012. "More Than Half B.C. Municipalities Vote against Pipeline Projects." *Vancouver Sun*. September 27.

Sklair, Leslie. 2002. *Globalization: Capitalism and Its Alternatives*. London: Oxford University Press.

Slocum, Tyson. 2001. "Electricity Utility Deregulation and the Myths of the Energy Crisis." *Bulletin of Science, Technology & Society* 21 (6): 473–81. http://dx.doi.org/10.1177/027046760102100605.

Smith, Graham. 2009. *Democratic Innovations: Designing Institutions for Citizen Participation*. Cambridge: Cambridge University Press. http://dx.doi.org/10.1017/CBO9780511609848.

SolarShare Co-operative. 2015. http://www.solarbonds.ca.

Sovacool, Benjamin. 2008. "Is the Danish Renewable Energy Model Replicable?" *Scitizen*. March 20. http://scitizen.com/future-energies/is-the-danish-renewable-energy-model-replicable-_a-14-1765.html.

–. 2010. "Critically Weighing the Costs and Benefits of a Nuclear Renaissance." *Journal of Integrative Environmental Sciences* 7 (2): 105–23. http://dx.doi.org/10.1080/1943815X.2010.485618.

–. 2011. "National Energy Governance in the United States." *Journal of World Energy Law & Business* 4 (2): 97–123. http://dx.doi.org/10.1093/jwelb/jwr005.

Stanfield, James, and Michael Carroll. 2009. "The Social Economics of Neoliberal Globalization." *Forum for Social Economics* 38 (1): 1–18. http://dx.doi.org/10.1007/s12143-008-9031-8.

Statistics Canada. 2002. *Electric Power Generation, Transmission and Distribution 2000*. Ottawa: Government of Canada Manufacturing and Energy Division.

–. 2010. *Canada at a Glance 2010.* Ottawa: Statistics Canada.

–. 2013a. *CANSIM Database Table 127-0009: Installed Capacity by Ownership and Source.* Ottawa: Statistics Canada.

–. 2013b. *CANSIM Database Table 127-0007: Electric Power Generation, by Class of Electricity Producer, Annual (Megawatt Hour).* Ottawa: Statistics Canada.

–. 2014. *CANSIM Database Table 127-0007: Electric Power Generation, by Class of Electricity Producer, Annual (Megawatt Hour).* Ottawa: Statistics Canada.

Stenkjaer, Nicolaj. 2008. *Wind Turbine Co-ops in Denmark.* Hurup Thy, Denmark: Nordic Folkecenter for Renewable Energy. http://www.folkecenter.net/gb/rd/wind-energy/48007/windturbinecoopsdk/.

Stephenson, Janet, Barry Barton, Gerry Carrington, Daniel Gnoth, Rob Lawson, and Paul Thorsnes. 2010. "Energy Cultures: A Framework for Understanding Energy Behaviours." *Energy Policy* 38 (10): 6120–29. http://dx.doi.org/10.1016/j.enpol.2010.05.069.

Steurer, Reinhard. 2013. "Disentangling Governance: A Synoptic View of Regulation by Government, Business and Civil Society." *Policy Sciences* 46 (4): 387–410. http://dx.doi.org/10.1007/s11077-013-9177-y.

Stocker, Thomas F., Dahe Qin, Gian-Kasper Plattner, Melinda Tignor, Simon K. Allen, Judith Boschung, Alexander Nauels, Yu Xia, Vincent Bex, and Pauline M. Midgley. 2013. *Climate Change 2013: The Physical Science Basis.* Working Group I contribution to the fifth assessment report of the Intergovernmental Panel on Climate Change – Abstract for Decision-Makers. Geneva: IPCC.

Stokes, Leah. 2013a. "Ontario's Backward Step on Renewable Energy." *Toronto Star.* July 22.

–. 2013b. "The Politics of Renewable Energy Policies: The Case of Feed-in Tariffs in Ontario, Canada." *Energy Policy* 56: 490–500. http://dx.doi.org/10.1016/j.enpol.2013.01.009.

Teeple, Gary. 2000. *Globalization and the Decline of Social Reform.* Toronto: Garamond Press.

Thomas, Stephen, and David Hall. 2006. *GATS and the Electricity and Water Sectors.* London: Public Services International Research Unit, University of Greenwich.

Thon, Scott. 2005. *Alberta Electricity Industry Restructuring: Implications for Reliability.* Calgary: Altalink Management.

Toke, David, Sylvia Breukers, and Maarten Wolsink. 2008. "Wind Power Deployment Outcomes: How Can We Account for the Differences?" *Renewable & Sustainable Energy Reviews* 12 (4): 1129–47. http://dx.doi.org/10.1016/j.rser.2006.10.021.

Trebilcock, Michael J., and Roy Hrab. 2003. *What Will Keep the Lights on in Ontario: Reponses to a Policy Short-Circuit.* Toronto: C.D. Howe Institute.

Trevena, Jack. 1976. *Prairie Co-operation: A Diary.* Saskatoon: Co-operative College of Canada.

Turnbull, Shann. 2007. "Analysing Network Governance of Public Assets." *Corporate Governance: An International Review* 15 (6): 1079–89.

Uluorta, Hasmet M. 2008. *The Social Economy: Working Alternatives in a Globalizing Era.* New York: Routledge.

United Nations. 2010. *Human Development Report 2010: The Real Wealth of Nations.* New York: Palgrave Macmillan.

–. 2013. *World Economic and Social Survey.* Geneva: United Nations.

United States Government Accountability Office (GAO). 2005. *Electricity Restructuring: Key Challenges Remain.* November. Washington, DC: GAO. http://www.gao.gov/assets/250/248509.pdf.

Vaillancourt, Yves. 2008. *Social Economy in the Co-construction of Public Policy.* Ottawa: Canadian Social Economy Hub.

Valentine, Scott Victor. 2010. "Canada's Constitutional Separation of (Wind) Power." *Energy Policy* 38 (4): 1918–30. http://dx.doi.org/10.1016/j.enpol.2009.11.072.

Vanderheiden, Steve. 2008. *Atmospheric Justice: A Political Theory of Climate Change.* New York: Oxford University Press. http://dx.doi.org/10.1093/acprof:oso/9780195334609.001.0001.

Victor, David, and Thomas C. Heller, eds. 2007. *The Political Economy of Power Sector Reform.* Cambridge: Cambridge University Press. http://dx.doi.org/10.1017/CBO9780511493287.

Vieta, Marcelo. 2010. "Beyond Capitalocentrism." *Affinities* 4 (1): 1–11.

Vikkelsø, Ann, Jens H.M. Larsen, and Hans Chr. Sørensen. 2003. *The Middelgrunden Offshore Windfarm: A Popular Initiative.* Copenhagen: Copenhagen Environment and Energy Office.

Walker, Gordon. 2008. "What Are the Barriers and Incentives for Community-Owned Means of Energy Production and Use?" *Energy Policy* 36 (12): 4401–5. http://dx.doi.org/10.1016/j.enpol.2008.09.032.

Walker, Gordon, and Patrick Devine-Wright. 2008. "Community Renewable Energy: What Should It Mean?" *Energy Policy* 36 (2): 497–500.

Walker, Gordon, Sue Hunter, Patrick Devine-Wright, Bob Evans, and Helen Fay. 2007. "Harnessing Community Energies: Explaining and Evaluating Community-Based Localism in Renewable Energy Policy in the UK." *Global Environmental Politics* 7 (2): 64–82. http://dx.doi.org/10.1162/glep.2007.7.2.64.

Warren, Charles R., and Malcolm McFadyen. 2010. "Does Community Ownership Affect Public Attitudes to Wind Energy? A Case Study from South-West Scotland." *Land Use Policy* 27 (2): 204–13. http://dx.doi.org/10.1016/j.landusepol.2008.12.010.

Weis, Tim. 2010. *Comparing U.S. and Canadian Investments in Sustainable Energy in 2010.* Ottawa: Pembina Institute.

Weis, Tim, Alison Bailie, Alex Doukas, and Greg Powell. 2009. *Green Power Programs in Canada 2007.* Ottawa: Pembina Institute.

Wilke, Marie. 2011. "Getting FIT for the WTO: Canadian Green Energy Support Under Scrutiny." *Bridges Trade BioRes Review* 5 (1): 2–4.

Wilkinson, Rorden, and Steve Hughes, eds. 2002. *Global Governance: Critical Perspectives.* London: Routledge. http://dx.doi.org/10.4324/9780203302804.

Williams, Chris. 2010. *Ecology and Socialism: Solutions to Capitalist Ecological Crisis.* Chicago: Haymarket Books.

Wood, Ellen Meiksins. 1995. *Democracy against Capitalism.* Cambridge: Cambridge University Press. http://dx.doi.org/10.1017/CBO9780511558344.

Wright, Eric Olin. 2010. *Envisioning Real Utopias.* London: Verso.

WTO (World Trade Organization). 2011. *Agreement on Government Procurement.* Geneva: WTO. https://www.wto.org/english/tratop_e/gproc_e/gp_gpa_e.htm.

–. 2013a. *Canada – Certain Measures Affecting the Renewable Energy Generation Sector* and *Measures Relating to the Feed-in Tariff Program.* Appellate Body reports. Disputes DS412 and DS426. New York: WTO.

–. 2013b. *Canada: Measures Relating to the Feed-in Tariff Program.* Dispute DS426. New York: WTO.

Yadoo, Annabel, and Heather Cruickshank 2010. "The Value of Cooperatives in Rural Electrification." *Energy Policy* 38 (6): 2941–47. http://dx.doi.org/10.1016/j.enpol.2010.01.031.

Žižek, Slavoj. 2009. *First as Tragedy, Then as Farce.* London: Verso.

Index

Note: FERC stands for Federal Energy Regulatory Commission; IPCC, for Intergovernmental Panel on Climate Change; NERC, for North America Electric Reliability Corporation; "(t)" after a page number indicates a table.

Aboriginal co-operatives, 16, 108–9, 131, 132, 134, 159, 223*n*12. *See also* First Nations

AGRIS Solar, 143–45(t), 166–67

Alberta: co-operative policy, 60–61, 120, 121–22, 190–91; co-operatives in, 22, 50(t), 58, 64, 116, 117–20, 123–26, 128, 140, 145(t), 148, 163–64, 169, 190, 197, 210(t); electricity sector, 9, 10, 11, 79(t), 81, 82, 83–84, 88, 91–92, 94(t), 99–100, 103(t), 207(t); energy policy, 94(t), 96, 98–100, 155; relationship between government and private business, 104–5

biomass power: about, 100, 103, 107, 112; and co-operatives, 1–2, 17, 22–23, 128, 132–33, 136, 142(t), 145(t)

Bookchin, Murray, 39–40

British Columbia: BC Hydro, 79(t), 81, 83, 94(t), 95(t), 100, 101, 101(t), 102; co-operative policy, 103, 103(t), 195;

co-operatives in, 24, 50(t), 51, 119(t), 123, 124(t), 126, 146, 146(t), 154, 170, 175, 222*n*1, 222*n*7; electricity sector, 10, 78, 79(t), 87, 88, 88(t), 91–92, 91(t), 92(t), 100–1, 103, 113, 120, 121, 197; energy policy, 12, 81–82, 93, 94(t), 96, 98, 100, 101(t), 102–3, 126, 195, 223*n*14

British Columbia Sustainable Energy Association (BCSEA), 170–71, 176

capitalism: criticism of, 33; definition, 33; and double movement, 35; and economic crisis, 33–34; and social economy, 34–35; and sustainability, 34–35; triple crisis, 5–6, 8, 9, 17–18, 33–36, 39. *See also* co-operatives, and capitalism

Central Alberta Rural Electrification Association (CAREA), 122, 124–26, 128, 149, 170. *See also* rural electrification associations (REAs)

climate change: and fossil fuels, 6–7;
greenhouse gas (GHG) emissions,
1, 6, 90(t), 185; IPCC, 6, 183–84;
reforms, 111–12; and renewables,
6–9, 183–84, 185
Cohen, Marjorie Griffin, 113–14
community: and co-operatives, 4,
131–32; definition, 129, 178;
development, 5, 13, 27, 42, 105,
122, 159, 167–68, 176, 192, 197;
empowerment, 5–6, 12–15, 46–47,
70–71; engagement, 10, 16, 23–24,
42, 46, 58, 147, 150–51, 161–62, 191,
193–94, 195; mobilization, 41–42, 57,
66, 116, 120–23, 142, 165–69, 176,
187–89, 190, 198, 201
Community Economic Development
Investment Funds (CEDIFs),
135, 170
community power: co-operatives, 9,
15–17, 169, 193; definition, 15–16,
165–66, 170–71, 178–79, 182, 222n5,
225; First Nations, 16; municipality,
16; non-profit, 17; and opposition
to projects, 140, 180–82; private for
profit, 17; and private sector, 177–78;
and public policy, 2–3, 22, 116, 119,
168–69, 171, 179–80, 190, 193–94;
and triple crisis, 9
Community Power Fund, 132–33, 141,
174, 222n2, 223n1
community energy. *See* community
power; electricity co-operatives
community feed-in tariff (COMFIT),
129–30, 130(t), 134–36, 137, 173.
See also feed-in tariff (FIT)
communitywash, 41–42, 115, 165–66,
176–80, 191–92
Conseil canadien de la cooperation et
de la mutualité, 62, 169, 170, 207.
See also Co-operatives and Mutuals
Canada (CMC)
continentalism, 10, 41, 82–83, 84–85,
86, 113–14
Co-operative Commonwealth
Federation (CCF), 15, 60

co-operative difference: as an
advantage, 51, 64, 66, 139–41, 191;
challenges, 57–60, 66, 154, 162,
192, 193–94; definition, 3, 224; and
education, 149–50; main elements,
51–52; and ownership, 147
co-operatives: and accountability,
13, 55–57; assets, 49–50; benefits,
12–13, 21–22, 36; and capitalism,
5, 40–41, 57–58, 66, 186, 193–94;
definition, 46, 48–49; and democracy,
3, 12, 13, 15, 55–56; drivers, 5, 9–10,
12, 21–22; economic growth, 13,
49–50, 60; and employment, 50,
141; and empowerment, 12; federal
initiatives, 61–62, 185; flexibility,
60–61; geographic distribution, 50–51;
and governance, 40, 40–43, 43(t),
56; history, 5, 12–13, 46, 47–48; and
innovation, 11, 30–31, 37, 56, 135, 138,
143, 195; interjurisdictional policy
learning, 171–73; and international
co-operative principles, 3–4, 52–53,
170; international membership, 13;
and legitimation, 14–15, 22, 24, 25, 61,
63, 64, 94, 168, 176, 179–80, 182–83,
192, 194–95, 201; limitations, 4, 14–15,
40–41, 57, 158; membership, 13, 47,
50–51, 55–56; and neoliberalism, 13,
14, 32–33, 40–41, 43, 43(t), 59–60,
64–65, 129, 184–85, 187, 193–94; nonprofit community associations, 23; and
privatization, 13, 41, 64; and profit,
3, 14–15, 53–55; provincial policies,
9, 23, 25, 61–63, 110–11, 119–23,
129–37, 190–91; and public policy,
13, 15, 22, 56, 66, 156, 165, 187; social
economy and, 13; and social justice,
14, 39–40; as a social movement, 13;
structure, 17, 22–23; survival rates,
55(t); and sustainability, 13, 14, 39–40,
137–38; as transformation agents,
14, 23–24, 36–37, 42–43, 45, 48; and
United Nations, 13, 53; volunteering,
32–33, 50, 58, 145–47, 158, 163–64.
See also electricity co-operatives

Co-operatives and Mutuals Canada (CMC), 62, 169, 170, 207. *See also* Conseil canadien de la cooperation et de la mutualité

co-optation of co-operatives, 40–41, 162. *See also* countervailing power

countervailing power, 40–43, 42(t), 45, 54, 59, 142–43, 165–66, 168, 194–95. *See also* co-optation of co-operatives

credit unions, 47, 187, 220*n*2, 227

Daly, Herman, 37–38

Danish Wind Turbine Owners Association, 20

democracy: and co-operatives, 5, 15, 141, 166; democratic legitimacy, 8; economic, 5, 34, 36, 58–59, 141, 166; liberal, 2; participatory, 2, 37, 38–39; substantive, 2, 36. *See also* governance

Denmark: electricity co-operatives, 3, 17–18, 18–21; energy sector, 19, 70–71, 106; nuclear power, 18–19, 106, 172; renewable energy, 7, 19(t), 20, 108

Devine-Wright, Patrick, 15–16

double movement, 35

eco-localism, 35, 37–40, 225

electricity: distributed generation, 11, 88; export, 77–80, 83–85, 114; governance, 3; lobbies, 12, 25, 70, 104–7, 114, 159, 164; ownership, 86, 87, 88, 88(t), 90, 92, 93–95, 102–3, 112; provincial profile of electricity generation, 78, 79(t), 87–93, 91–92; price, 87, 89, 89(t), 112–14; public sector involvement, 10, 87–88, 89; renewables, 87, 88–90, 103–4; sustainability, 12; trade, 78–79, 84–85; transformation, 25, 67; and US, 68, 77–84. *See also* electricity generation; electricity sector restructuring; energy sector

electricity co-operatives: from 1940 to 1990, 22, 116–18, 119–23; from 1990 to 2013, 117–19, 123–29; access to capital, 24; assets, ownership and control, 4, 49–50, 118, 120, 142–47, 148–52, 161–64, 191–92; challenges, 23, 24–25, 139–41, 152–61; and community, 9, 158–61, 167–68; consumer co-operatives, 22–23, 48–49, 49(t), 117–18, 127–29; and democratic governance and accountability, 25, 55–57, 58–60, 193–94; in Denmark, 3, 17–18, 18–21; development stages, 141–42; distribution co-operatives, 22, 117, 119, 123–26, 163; education, 23–24, 149–52, 176, 191; financing, 4, 24–25, 156–58, 129; generation co-operatives, 22–23, 117, 126–27, 140, 163; in Germany, 3, 21, 24–25, 220*n*1; grid and site access, 23, 153–56; historic development, 22, 47–48, 116–19, 119–20, 122, 123, 129, 183; information and capacity building, 24, 57; and local communities, 4, 21–22, 141–42, 147; membership, 48–49, 143, 148–49; multi-stakeholder/ solidarity co-operatives, 49; network co-operatives, 118, 127–29; networking, 22, 165–66, 169–73, 192–93; and the private sector, 24, 119, 129, 140, 152–53, 156–57, 161, 176–78, 187; profit, 53–55, 141, 147, 149, 157, 179; and public policy, 119, 120–23, 129–37, 165, 168–69, 190–91; retail, 22, 47, 49, 117–18, 123, 128–29, 140, 142(t), 143–44, 163–64, 183, 190, 191–92; and sustainability, 1–2, 3, 14, 22–23, 140–41, 166–69, 170, 187, 189, 194–98; volunteering, 23, 32–33, 50, 58, 145–47, 158, 163–64; worker co-operatives, 22, 40, 48–49, 49(t), 61, 62, 118, 127–29. *See also* community; co-operatives

electricity generation, 78, 79(t), 87–93, 91–92. *See also* electricity sector restructuring; fossil fuels; renewable energy

electricity sector restructuring:
Alberta, 10, 94(t), 98, 99–100; British
Columbia, 10, 100–1; challenges,
10–11, 82–84, 109–11, 114; in Chile,
71–72; and competition, 75–77, 82;
and co-operatives, 4, 25, 86–87, 91,
102–3, 139–40, 184–85; deregulation,
10, 22, 31–32, 33–34, 68, 72, 76,
98–99, 112, 114, 117–18, 124–25,
226; drawbacks, 11–12, 74–77, 114;
drivers, 11, 12, 67–68, 70, 84, 86–87,
93, 96–97; governance, 70–71, 86,
93–95, 96(t), 113–14; independent
power producers (IPPs), 25, 69, 75,
86–87, 126; and innovation, 69–70,
96–97; international pressure, 67,
77–85; and neoliberalism, 67–68,
69, 76, 77, 93, 111–12, 114, 115–16,
195; Nova Scotia, 82, 93, 95(t), 97;
Ontario, 10, 81–82, 94(t), 97–100,
107–11; piecemeal restructuring,
10, 100–2; privatization, 10, 12, 69,
75–77, 86–87, 93; and provinces, 10,
79–80, 86, 91–92, 93–96; in Quebec,
102; and renewable electricity, 11, 70,
76, 90–91, 102–4, 107–11, 111–14;
technical limitations, 74–75; in the
UK, 68, 69–70, 71– 73, 72, 73–74
energy efficiency, 15–17, 18–20, 70–71,
91, 100, 101, 118, 128, 194
energy lobbies, 25, 70, 105–7, 114,
159, 164
energy sector: and electricity grids,
82–83; fossil fuels, 8, 87, 90,
92; renewable energy, 8, 89–92;
structure, 68–69
energy security, 72, 78, 90, 114, 221*n*3

Federal Energy Regulatory Commission
(FERC), 77, 80–82, 82–83, 84–85,
221*n*2
Federated Co-operatives Limited (FCL),
47, 49, 54
Federation of Community Power
Co-operatives, 23, 127, 152–53,
155–56, 166, 169, 173–74, 219*n*15

feed-in tariff (FIT), 23, 107–10, 127,
129, 130–37, 143–44, 145(t), 155–56,
158–59, 163, 173, 176, 182, 215–17,
223*n*10
First Nations: electricity co-operatives,
15, 16, 108, 109, 110, 131–32, 133,
135, 137, 144, 155, 163; renewable
energy, 182. *See also* Aboriginal
co-operatives
Forsyth, Tim, 32
fossil fuels: coal, 6–7, 8, 9, 69–70, 73,
78, 84, 87, 89–90, 91(t), 92, 96(t),
105–6, 107, 108, 112, 114, 119, 172,
173, 184–85, 197, 221*n*1, 221*n*4;
natural gas, 6–7, 8, 17–18, 19–20,
24, 64, 77–78, 89, 90(t), 91(t), 92,
96(t), 105, 112, 121–23, 125–26, 146,
169, 185, 190, 194, 221*n*4; oil, 10,
54, 77–78, 83–84, 90(t), 91(t), 96(t),
105, 185
Fourth Pig Worker Co-op, 128–29
free trade. *See* North America Free
Trade Agreement (NAFTA); trade
agreements
Fung, Archon, 36, 40–42, 166, 195–96,
218*n*1

Germany: electricity co-operatives,
3, 21, 24–25, 220*n*1; energy sector,
18, 70–71; and nuclear energy, 106;
renewable energy, 7, 19(t), 108, 221*n*7
Gipe, Paul, 59, 111, 174
globalization, 31, 48, 53, 187
governance: definition, 30; empowered
participatory governance (EPG),
2, 6, 28, 40–43, 43(t), 49, 189, 191,
225; energy governance, 70–71;
localized governance, 37; neoliberal
governance, 30–33, 33, 34, 36, 59–60;
sustainable governance, 28–29. *See
also* democracy
Gramsci, Antonio, 35–36
Green Energy Act Alliance, 133, 141,
152–53, 171, 173, 174
Green Energy and Green Economy
Act, 107–11, 130–31, 133, 143–44,

152–53, 159–61, 169–70, 173–74, 178, 182, 185–86, 195, 219n12. *See also* Ontario, co-operative policy
green political theory, 38–39, 226
greenwashing, 111–14, 201, 226
greenshifting, 186

Harvey, David, 33–34
hydropower: about, 8, 10, 11, 77, 78, 81–82, 84, 87–88, 89–90, 91(t), 93, 94(t)–95(t), 96(t), 98–99, 100–2, 105, 106–7, 109(t), 112, 114, 151–52, 184, 196–97, 214(t), 219n8, 221n4, 221n6, 221n5; co-operatives, 2, 126, 133, 142(t), 145(t)–46(t); Hydro-Québec, 120–21, 133, 142(t), 143, 146(t), 157

independent power producers (IPPs), 69, 25, 69, 75, 86–87, 95(t), 97, 100, 101(t), 108, 126, 134, 164, 179. *See also* electricity sector restructuring, independent power producers (IPPs)
International Co-operative Alliance, 46, 52–53
International Energy Agency (IEA), 84
investor-owned utilities (IOUs), 122–24, 142(t), 149, 164, 169. *See also* utilities

Laidlaw, Alexander Fraser, 45
Lamèque Renewable Energy Co-operative, 146–47, 147(t), 152, 162
local trap, 39–40
localism: about, 11, 37–38, 39–40, 58–59; and co-operatives, 4, 188; and sustainability, 37–40. *See also* eco-localism

MacPherson, Ian, 1–2, 3–4, 5, 47, 48–49, 53, 64–65
Manitoba: co-operative policy in, 62–64, 208(t); co-operatives in, 50(t), 119, 124(t), 196; electricity sector in, 78, 79(t), 88, 91–92, 103(t), 120, 197, 214(t); energy policy in, 81, 87, 95(t)
market competition, 11–12, 25, 31, 46,

57, 67, 70, 72–73, 75–77, 79, 82, 84, 94(t), 95(t), 96–100, 107
Marx, Karl, 3, 33, 34–35, 219n1
McMurtry, John J., 2, 14, 15, 22, 35, 36
Mendell, Marguerite, 14, 133, 165
microFIT, 23, 108–9, 127, 143–44, 145(t), 215(t)–17(t). *See also* feed-in tariff (FIT)
Mountain Equipment Co-operative, 49, 51
multi-stakeholder co-operatives, 48–49, 49(t). *See also* solidarity co-operatives

National Energy Board (NEB), 77, 80–81, 83–84
NIMBY ("not in my back yard"), 22, 141, 172, 176, 180–82
North America Electric Reliability Corporation (NERC), 77, 80–81, 82–83, 84–85
North America Free Trade Agreement (NAFTA), 79–81, 82–83, 198, 199
neoliberalism: in Canada, 31, 64–65, 186, 218n4, 226; definition, 8–9, 26, 31–32, 111. *See also* co-operatives, and neoliberalism; privatization, and neoliberalism
new co-operativism, 14, 57–58, 60. *See also* solidarity economy
New Brunswick: co-operative policy in, 23, 129, 130(t), 133, 136–37, 165, 190, 209(t); co-operatives in, 50(t), 119(t), 119, 124(t), 153, 197; electricity sector in, 79(t), 80, 81, 88(t), 91(t), 91–92, 92(t), 102, 103(t), 112; energy policy in, 95(t), 107
Newfoundland and Labrador: co-operative policy in, 63; co-operatives in, 51(t); electricity sector in, 88(t), 91–92, 91(t), 92(t), 103(t), 209(t); energy policy in, 87, 95(t)
non-profit organizations, 15, 17, 19–20, 22–23, 32–33, 53–54, 127–28, 132–33, 137, 149–51, 170, 187, 188–89, 193, 207(t), 220n1, 224. *See also* social economy

Northwest Territories: co-operative policy, 209(t), 223*n*12; co-operatives in, 51(t); electricity sector in, 92(t)
Nova Scotia: co-operative policy, 23, 61, 62–64, 129, 130(t), 133, 134–36, 137, 165, 168, 170, 175, 179, 190, 208(t); co-operatives in, 47, 50(t), 55, 57, 119, 124(t), 126, 156–57, 175–76, 181, 190, 197; electricity sector, 9, 11, 12, 79(t), 82, 86, 87–88, 91–92, 93, 112, 156, 181; energy policy, 95(t), 96–97, 100, 107, 112, 221*n*1; Nova Scotia Power, 93, 97
Nova Scotia Sustainable Energy Association (NovSEA), 134, 175–76
nuclear power: and co-operatives, 23, 24, 105, 153, 155, 158; employment, 105; industry, 7, 8, 11, 89, 90(t), 91(t), 96(t), 112; lobby, 25, 70, 105–7, 114, 159, 164; as renewable energy, 87
Nunavut: co-operative policy in, 209(t); co-operatives in, 51(t); electricity sector in, 91–92, 92(t)

Ontario: co-operative policy, 23, 107–11, 126, 127, 129, 130–33, 130(t), 143–44, 152–53, 157–58, 159–60, 163, 165, 169–74, 178, 182, 185–86, 190, 195, 219*n*12; co-operatives in, 2, 22, 50, 55, 56, 60–61, 115–19, 121; 123, 124(t), 126–27, 128, 139, 142, 150–51, 154–55, 163, 173–76, 177, 189, 190–91; electricity sector, 19(t), 23, 74, 75, 77, 78, 79(t), 80–83, 88(t), 91(t), 91–92, 92(t), 98–100, 103(t), 105–6, 112–13, 120, 178–79, 197, 222*nn*3–4; energy policy, 9, 10, 12, 19(t), 70, 93, 94(t), 96, 97–100, 103, 135, 136, 137, 182, 185, 208(t), 215(t), 219*n*12, 221*n*1; Ontario Power Generation, 98–99, 106. *See also* Ontario Sustainable Energy Association (OSEA); Pukwis Community Wind Park; Toronto Renewable Energy Co-operative (TREC); Val-Éo Co-operative

Ontario Sustainable Energy Association (OSEA), 170–71, 173–76, 178–79, 190
open access transmission tariffs (OATTs), 81–82, 93
opposition to energy projects: about, 7, 109–10, 133; co-operative effect on, 139–41, 168–69, 172, 179–80, 180–82, 188–89; NIMBYism, 176, 180–82
Organisation for Economic Co-operation and Development (OECD), 76–77
organizational practice, 171
Ostrom, Elinor, 6, 7–8, 29–30, 37, 54–55, 56, 187
Ottawa Renewable Energy Co-operative, 24, 172, 175

partnership: cross-sector partnership (CSPs), 32; First Nations, 163; municipalities, 140, 163; pressure to partner, 163–64; with private, 24, 140, 154–55, 161–63; ownership, 24, 163; trade-offs, 140, 161–64
Pateman, Carole, 2, 38, 187
Peace Energy Co-operative, 24, 147, 175, 191
Polanyi, Karl, 34–35, 46
power sector restructuring. *See* electricity sector restructuring
Prince Edward Island: co-operatives in, 47, 51(t), 119(t), 124(t), 209(t); electricity sector in, 79(t), 88, 88(t), 91–92, 91(t), 92(t), 103(t), 120; energy policy in, 95(t)
privatization: about, 8, 10, 12, 69–70, 112, 225; Canada, 11, 86, 93, 94; and capitalism, 33; and neoliberalism, 8–9, 31–32, 69–70. *See also* co-operatives, and privatization
public policy: constraints, 104–7, 119, 152–53, 159–60, 165, 184–85, 201; importance of, 21, 23, 26, 30–31, 46–47, 59, 122, 129, 138, 167–69, 171; trends, 18–19, 21, 60–62, 116–37, 165, 176, 183, 195–97. *See also* community power, and public policy;

co-operatives, interjurisdictional
policy learning; co-operatives,
and public policy; electricity
co-operatives, and public policy
Pukwis Community Wind Park, 152,
158–61

Quarter, Jack, 13–14
Quebec: co-operative policy in, 23,
62–63, 120–21, 122, 129, 130(t),
133–34, 137, 165, 190, 208(t),
223*n*13; co-operatives in, 22–23, 47,
49, 50, 54, 55(t), 116, 119–20, 123,
124(t), 126, 127, 142(t), 143, 146(t),
157, 167–68, 177–78, 196, 209(t),
214(t), 218*n*5; electricity sector in, 8,
78, 79(t), 81–82, 88–89, 91–92, 97,
100, 102, 103(t), 105, 113–14; energy
policy in, 87, 94(t), 102, 190

renewable energy: challenges, 7–8,
12, 30–31, 88, 105, 111–14, 152–61,
176–77, 180–82, 186, 188; and
co-operatives, 3, 9, 17, 22, 26,
118, 119, 123–30, 139–41, 147,
149–52, 166–69, 170–71, 174, 177,
179–80, 188–89, 191; distributed
generation, 11, 70; global trend, 184;
privatization, 93, 102–4, 112, 139,
176, 194–95, 196. *See also* biomass
power; hydropower; solar power;
tidal power; wind power
renewable energy policy: by Canadian
federal government, 14, 88, 184–86;
by Canadian provinces, 9, 10,
93–102, 107–11, 127, 129–37,
152–53, 159–60, 168–69, 171,
190–91, 215(t); in the US, 72
Renewable Energy Standard Offer
Program (RESOP), 107–8, 109(t),
127, 152–53, 159–60, 174, 215(t). *See
also* Ontario, co-operative policy
Rosenau, James, 30
rural electricification associations
(REAs), 123–26, 128, 143, 148–49,
169–70, 190

Saskatchewan: co-operative policy
in, 64; co-operatives in, 47, 50(t),
56, 119(t), 124(t), 197; electricity
sector in, 9, 79(t), 81, 91–92, 91(t),
92(t), 102, 103(t); energy policy in,
95(t), 207(t)
Scheer, Hermann, 11, 67, 70
Schumacher, Ernst F., 37, 38–39
Sen, Amartya, 5–6
social economy: about, 1–2, 70–71, 226;
and co-operatives, 13, 39–40
social enterprises, 53–54, 193–94
solar power: about, 9, 74, 89–90,
91(t), 96(t), 100, 102, 103–4,
106–8, 109(t), 110, 112, 130–33,
134, 197; and co-operatives, 16,
22, 123, 127, 128, 133, 140, 142(t),
143–44, 145(t), 147, 150, 163, 166,
173–75, 181, 192
SolarShare, 143–44, 145(t), 150,
173–74, 191–92, 222*n*9
solidarity co-operatives, 43(t), 49,
60–61, 65–66, 187
solidarity economy, 43(t), 65–66, 194,
227. *See also* social economy
Spark Energy Co-operative, 128
sufficiency, 15–16, 37, 84–85, 95(t)
sustainability: and co-operatives, 1–2,
3, 11, 13, 14, 39–40, 137–38, 140–41,
189; definition, 28–29
Sustainability Solutions Worker Co-op,
128–29

tidal power: about, 9, 89, 90(t), 91(t),
96(t), 103–4, 112, 130(t), 134, 135,
184, 195; and co-operatives, 1–2
Toronto Renewable Energy
Co-operative (TREC), 22–23,
127–29, 142–43, 144, 149–50, 153,
155, 163, 166, 169–72, 173–75, 178,
189, 191, 200, 223*n*1
trade agreements: General Agreement
on Trade in Services (GATS),
79; North American Free Trade
Agreement (NAFTA), 79–81,
82–83, 198, 199. *See also* Federal

Energy Regulatory Commission (FERC) triple crisis, 5–6, 8, 9, 17–18, 33–36, 39. *See also* capitalism

United States: California electricity crisis, 10–11, 33–34, 68–70, 72, 74, 75–76; continentalism, 67–68, 77–78, 80–81, 82–85; electricity sector restructuring, 78; FERC, 77, 80–81, 82–83, 84–85, 221n2; NERC, 77, 80–81, 82–83, 84–85
utilities: electricity, 1, 9–12, 69, 72–73, 75, 82, 87–88, 91, 93, 94(t), 96–97, 100–2, 109, 112–14, 195–97; integrated, 88, 93, 94(t), 97, 109; and privatization, 9–10, 31, 47, 69, 72, 86, 93, 96–97, 112–13, 123; public, 69, 75, 87, 88, 91, 101–2, 114, 195–97; transformation of, 1, 9–12, 69, 72–73, 82, 93, 100. *See also* investor-owned utilities (IOUs)

Val-Éo Co-operative, 133–34, 143, 146(t), 163, 172–73
Vancouver Renewable Energy Co-operative, 129

Vieta, Marcelo, 14, 57–58, 60
Viridian Energy Co-operative, 128

Walker, Gordon, 16, 162, 169
wind power: about, 72, 74, 89, 90(t), 91(t), 94(t), 104(t), 108, 109-10, 109(t), 112, 135, 147, 153, 156, 158–59, 166, 167, 177–79, 184; and co-operatives, 15–16, 22–23, 24, 32–33, 102–3, 111, 127, 128, 130(t), 132–33, 136–37, 140–47, 151, 155, 158–59, 162–63, 168–69, 180, 181; Denmark, 17–18, 20–21, 89, 90, 103, 104(t), 112, 163–64, 172, 191, 199
WindShare, 127, 129, 140, 142–43, 145(t), 147, 149–50, 151, 153, 163, 171–72, 173–75, 177, 182, 183, 189, 200
World Bank, 26, 73
World Wind Energy Association Community Power Working Group, 179
Wright, Eric Olin, 2, 12, 29–30, 36, 38, 40–42, 166, 187, 195–96, 218n1

Yukon: co-operatives in, 47, 51(t), 209(t); electricity sector in, 91(t), 91–92, 92(t), 102, 103(t)